DIGITAL CCTV

DIGITAL CCTV

A Security Professional's Guide

Emily Harwood

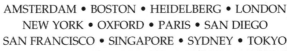

AMSTERDAM • BOSTON • HEIDELBERG • LONDON
NEW YORK • OXFORD • PARIS • SAN DIEGO
SAN FRANCISCO • SINGAPORE • SYDNEY • TOKYO
Butterworth-Heinemann is an imprint of Elsevier

ELSEVIER

Acquisitions Editor: Pamela Chester
Assistant Editor: Kelly Weaver
Marketing Manager: Marissa Hederson
Cover Designer: Eric Decicco

Elsevier Academic Press
30 Corporate Drive, Suite 400, Burlington, MA 01803, USA
525 B Street, Suite 1900, San Diego, California 92101-4495, USA
84 Theobald's Road, London WC1X 8RR, UK

Library of Congress Cataloging-in-Publication Data
Harwood, Emily.
    Digital CCTV / Emily Harwood.
      p. cm.
    Includes bibliographical references and index.
    ISBN-13: 978-0-7506-7745-5 (alk. paper)
    ISBN-10: 0-7506-7745-7 (alk. paper)
    1. Closed-circuit television. 2. Digital video. I. Title. II. Title: Digital closed
circuit television.
    TK6680.H356 2007
    384.55'6—dc22

                                                                        2007004218

British Library Cataloguing in Publication Data
A catalogue record for this book is available from the British Library

ISBN 13: 978-0-7506-7745-5
ISBN 10: 0-7506-7745-7

For all information on all Elsevier Academic Press publications
visit our Web site at www.books.elsevier.com

Table of Contents

About the Book

WHO IS THIS BOOK FOR?

CEOs
Security Managers
Directors of Security
Loss Prevention Managers
CCTV Product Manufacturers
Electronic Security Sales
    Personnel
Electronic Security
    Manufacturers
    Representatives

IT Managers
Security Systems Integrators
Electronic Security Installers
Security Dealers
Security Consultants
Architects and Engineers
Specifiers

WHAT IS THE PURPOSE OF THIS BOOK?

The purpose of this book is to provide you, the reader, with the information you need to interpret what is behind all of the technology smoke and acronym mirrors surrounding digital video technology enabling you to better understand today's new digital products. At last you will be able to answer puzzling questions about digital technology like how much storage space and bandwidth are necessary to handle digital video at specific quality levels and image rates.

This book provides practical information about how digital video works, how digital video is stored and transmitted, what digital systems can and cannot accomplish, and what to expect from digital video equipment in modern CCTV systems.

An explanation of digital video and compressed digital video is provided, and the distinction between raw digital and compressed digital video is explained. After a basic understanding of how these differences affect the video image is reached by the reader, things like picture quality, resolution, and evidentiary use of digital video will be easier to comprehend. Compression variables such as lossless and lossy will be explained by reviewing Huffman and Run Length Encoding (RLE). A review of JPEG, motion JPEG, MPEG, and wavelet compression schemes, among others, will also be provided.

Introduction

Growth naturally stimulates change, and CCTV technology has been no exception. A system that once merely required cameras, cabling, and video monitors has now become a complex electronic configuration of equipment intertwined with both computer and telecommunications technologies. This dramatic change is directly related to the introduction of digital technology. Why do we need to understand how digital technology works, and what does it have to do with the future of security? It's simple—the newest revolution in technology is pervasive computing. Computers are or soon will be everywhere, linked to everything, and everything will be connected by the Internet—including security systems.

Upheavals within the electronics industry have been persistent and are well known. For example, most everyone remembers how eight track players were relegated to the trash heap without so much as a backward glance. Phonograph records were shut out by compact discs and the consumer VCR has virtually been replaced by the DVD player. In the security industry, the revolution from analog to digital is similar to these earlier advancements and will probably be looked at with the same amount of disdain regarding archaic processes of the past. Digital technology is exploding around us, yet a large amount of industry professionals are still looking for a comprehensive explanation of digital video as a security technology.

Security professionals today understand how the components of a CCTV system work. They know the applications, limits, strengths, weaknesses, and relative costs of lenses, cameras, camera mounts, pan/tilt units, and housings. Such knowledge enables professionals to design systems and to select from a myriad of products just the right components, resulting in a CCTV system that will meet customer performance requirements and budgets.

There is, however, a concern that digital CCTV equipment concepts have not been adequately explained. The reality is that digital technology is much more than a trend and requires a rather extensive learning process if one can intelligently buy, sell, install, or recommend digital video products. In today's environment, it is essential for the security professional to know how the Internet works and how LANs and WANs function in relation to the World Wide Web.

WHY SWITCH TO DIGITAL?

There are many reasons to make the switch to digital for security surveillance and recording applications. Probably the strongest reason is that digital information can be stored and retrieved with virtually no degradation, meaning that with digital images, copies are as good as the originals. When a digital recording is copied, it is a clone, not a replica.

Digital information is not subject to the noise problems that degrade analog information as quickly as it is stored, retrieved, and duplicated. There are no amplifiers to introduce distortions and noise to a digital signal. When transmitting images, a digital system reduces noise over successive transmissions because small variations in the signal are rounded off to the nearest level. Analog transmission systems must filter out the noise, but the filter itself can sometimes be a source of noise.

In some ways, digital information outperforms analog information. For example, digital music from a CD has a much wider dynamic range (very quiet to very loud) than analog music from a tape or a record. With all of the advancements available in digital technology, it is not as "perfect" as analog video and does present a variety of new problems in transmission and storage. Because digital video consists of large amounts of data, it must be compressed, in most cases, to be useful. Compression discards a significant amount of the original information and results in a new kind of degradation called "artifacts". This discarding of information by compression techniques has raised questions about whether digital video or compressed digital video can be used as evidence in a court case.

WHAT ELSE CAN DIGITAL DO FOR VIDEO?

As an added bonus, most digital video systems permit the manipulation of devices from a location off-site. Pan/tilt/zoom features on cameras can be controlled allowing an enhanced portrayal of events as they occur; motorized gates, electric door locks, lights, and environmental controls can be remotely activated as well. With these features, approved access can be controlled off-site and the expensive misuse of utilities can be monitored and corrected instantly.

Digital images of a crime or a hazardous situation of some type can be transmitted over a wireless local area network to first responders for evaluation. The use of an IP network to transmit these images can allow access to the system from any device with an Internet connection and proper authorization for access.

The benefits of digital video transmission technology in the security arena are limitless. Intelligence can be programmed into a digital system so that it will "look" for specific analogies and respond in some manner. Digital video systems can automatically zoom in on individual faces to improve or verify identification. Video verification of events is immediate—intruders can be positively identified, false alarms eliminated, and facility management improved—all with one system.

Many other intelligent operations can be integrated with a digital system to expand its functionality. Networked video systems permit remote surveillance via WAN/LAN and Internet infrastructures. With an open-architecture design, networked digital systems can provide easy integration with other technologies including access control, facial recognition, points of sale, and database systems.

There are significant economic considerations for using digital technology. Digital circuits can be manufactured for less money than analog circuits due to the fact that analog circuits require resistors, capacitors, diodes, chokes, transformers, and other discreet components to make things work. Digital circuits also use many of these components but they are typically much smaller, surface mount components and not as many are needed since IC (Integrated Circuit) chips replace many of them. The largest portions of digital circuits are simple on/off transistor switches that can easily be applied to integrated circuits in large quantities. Also, integrated circuits can be mass-produced, which drives down costs.

In most cases you will obtain more performance per dollar spent with digital than with analog video. Once video has been digitized, it can be used virtually anywhere in the world and with the aid of communications links like telephone, Internet, and various wireless technologies, it can be transmitted anywhere in the world as well. TCP/IP transmittal of surveillance video is now a viable and economical mode of remote monitoring of multiple locations.

Unlike digital signals, which are composed of ones and zeros and can pass through a wire or be recorded to tape with absolutely no change, analog signals are composed of information, which will change slightly every time it goes through a wire or gets recorded to tape. The ultimate quality of an analog process is not inherently inferior; it is very difficult to keep the original quality through the entire production pipeline.

DIGITAL TECHNOLOGY REDUCES MANPOWER REQUIREMENTS

Until recently, video surveillance technology has relied on human operators for detecting breaches and facilitating appropriate responses, making the surveillance only as effective as the operator. Because advances in technology have made it possible to integrate more cameras and send images virtually anywhere in the world, there is a growing potential for an overload of information resulting in operational inefficiency. For a large surveillance system with hundreds of cameras, the fatigue factor is extreme. These adverse conditions can be overcome by utilizing new advancements in the technology of video surveillance.

Software that intelligently monitors images and automatically detects potential security threats changes the dynamics of video monitoring for security. Today's digital video surveillance systems are much more than camera eyes that view and record the scenes around them. Surveillance systems now analyze and make decisions about the images they are viewing based on the confirmation or violation of preset protocols. The system immediately relays information to human operators (or in some cases to other security or operational systems) for immediate action. The resulting investigation of suspicious incidents help operators makes the right decision, on time.

How does it work? Analytics transform video into security information. Software programs that utilize complex mathematical algorithms to analyze scenes in a camera view are designed to detect predetermined behaviors such as someone lying on the floor, erratic movements, people or cars converging on each other, a person or vehicle staying in one place for an extended period, a person or vehicle traveling against the normal flow, objects newly appearing on the scene—the list continues to grow. These types of programs tremendously increase a security officer's efficiency.

THE ENIGMA OF DIGITAL VIDEO

Over the last few years, there has been more and more news media coverage on the subject of video for security in the US. The use of CCTV for surveillance is by no means new, but from some news clips, you might think it is the latest invention in crime detection and investigation. The community inside of the security industry knows how prevalent the use of video is and that the new benefits arriving with digital advancements are almost exponential. For outsiders, the news is not as common. In fact CCTV, digital video surveillance and intelligent video solutions cover such a wide range of relevance that these subjects almost always have to be covered from the very beginning to the present.

The adage "time waits for no man" could not be more applicable than in the world of digital technology. Even as these words are being written, new developments are underway all over the world, which will continue to contribute additional cost effective, efficient alternatives for the compression and transmission of video, audio, and data.

1

We Live in an Analog World

The security world is well acquainted with the term Closed Circuit Television (CCTV), which is a visual surveillance technology designed for monitoring a variety of environments and activities. CCTV systems are used in applications such as monitoring public areas for violent actions, vandalism, theft, and unlawful entry, both indoors and out. CCTV recordings are used to obtain and provide evidence for criminal and other investigations; they are sometimes disclosed to the media in the hopes of gaining information about images of a suspect or suspects caught in or near a crime scene.

The term Closed Circuit Television can be misleading, as the word television actually means to see at a distance, which implies broadcast. If public broadcast is not the intent, CCTV is the correct terminology, as it is not a system for broadcast to the public in general. Unlike television that is used for public entertainment, a CCTV system is closed and all its elements are directly connected either by hardwire methods or wireless technologies.

1

Wireless analog devices typically use line of sight radio frequency that can usually only be transmitted for short distances. Some newer technologies, however, can transmit for several miles. This means that the transmitted video can only be viewed with the proper equipment set to the proper frequency. While the signal could be intercepted, it is still considered a closed circuit since it is not used for a multi-point broadcast such as cable TV.

It is important to review some of the key concepts related to analog video in order to have an understanding of how these concepts play a role in digital video. The word video comes from the Latin verb *videre*, "to see", and is commonly used when referring to devices such as video monitors or video recorders. In this book, video will also refer to the actual product of the technology, that is to say, the image produced. The purpose of this first chapter is to acquaint the reader with the basics of analog video as it is normally used in a security function. For some readers, this chapter will merely be a review of basic analog video theory. For others, it may introduce or explain various concepts in enhanced detail. For a number of readers, it will be a primer of video concepts.

HOW AN ANALOG VIDEO IMAGE IS GENERATED

We live in an analog world, and vision is an analog function. Waves and electromagnetic fields are analog, meaning they are continuous signals capable of smooth fluctuation. Electric current, characterized by its flowing current, is also analog. Electricity is a current of electrons with either a direct flow or current called DC or an alternating flow or current called AC. In an analog CCTV system, an analog camera "sees" an event, which it turns into an electronic signal. It then transmits the signal over some type of medium and the signal terminates at a display or recording mechanism. In the United States, a video image is made up of 525 horizontal lines, according to the NTSC standard. NTSC stands for National Television System Committee, which devised the NTSC television broadcast system in 1953. One still picture or frame of video consists of two scans containing 525 alternate horizontal lines that are produced by a ray of electrons. The camera and picture tube first scan 262.5 odd numbered lines, and then the picture is scanned again to form 262.5 even numbered lines. Each half of the frame or 262.5 lines is one "field" of video. After the ray or beam of electrons writes the lines one at a time onto a picture tube, one frame of video is created.

This operation of assimilating a picture, translating that picture for transmission, and then scanning that same picture at the receiving location results in the successful transmission of one full frame of video. The time involved in this operation from beginning to end is the "update" or "refresh" rate. After the process is repeated thirty times, the illusion of motion is created. This is the same principle used for creating flipbooks—you quickly flip through to see a moving picture. Cartoons that are drawn and rapidly displayed one picture at a time use the same technique to create perceived motion. Each of the 30 frames is a still image of a scene, and by slightly changing something in each scene, the viewer will perceive a progressively changing or moving image.

Analog video is comprised of continuously varying voltage levels that are proportional to (analogous to or the same as) the continuously varying light levels in the real world. When we refer to electronics in relation to video, we are referring to the use of current and voltage to carry electric signals modified to represent information. If we can convert picture information into electronic or radio signals, we can send it virtually anywhere in the world with the right transmission system.

A very simple explanation of video transfer goes something like this: imagine that the camera is the eye of the system and its function is to make its view (the image) available in an electronic format of impulses. These impulses are then propelled along wires, cables, or microwaves via voltage, which is the pressure or electromotive force that compels electrical charges to move from negative to positive. The result is the transfer of video information from the camera to its ultimate destination. See Figure 1-1.

Wires and certain other parts of circuits are made of materials called conductors. These conduits carry the electric currents. Wireless transmission technology will be discussed in a later chapter. For now, let's just acknowledge that video signals can be transmitted without the benefit of wires as conductors. Electromagnetic waves are unique forms of energy, known as radiant energy. They

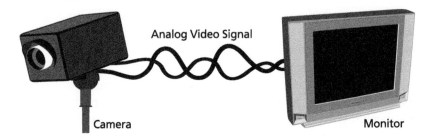

Figure 1-1   Video Transfer

are created when electrically charged particles, such as electrons, are made to move. As the charged particles move, they generate fields of electrical and magnetic energy. These two forms of energy radiate from the particles as electromagnetic waves.

Energy is a property of many substances and is associated with heat, light, electricity, mechanical motion, and sound. Energy is transferred in many ways. In physics, the transfer of energy by some form of regular vibration or oscillatory movement, like an electromotive force, is called a wave. An electromagnetic wave consists of two primary components—an electric field and a magnetic field. The electric field results from the force of voltage, and the magnetic field results from the flow of current. Although electromagnetic fields that are radiated are commonly considered to be waves, under certain circumstances their behavior makes them appear to have some of the properties of particles. In general, however, it is easier to picture electromagnetic radiation in space as horizontal and vertical lines of force oriented at right angles to each other.

Frequency is the measure of the number of waves that pass through a fixed point in a specified period of time—often measured as cycles per second. One cycle per second is called a Hertz (Hz), one thousand is called a kiloHertz (KHz), and one million is called a megaHertz (MHz). The amplitude of a wave is defined as the measurement from its crest to its trough. The distance between consecutive crests or troughs is the wavelength. The frequency of a wave is equal to the number of crests (or troughs) that pass a fixed point per unit of time. The smaller the wavelength is, the greater the frequency is. See Figure 1-2.

Properly terminated video signals have amplitude of one volt peak-to-peak. This means the total voltage produced is one volt from

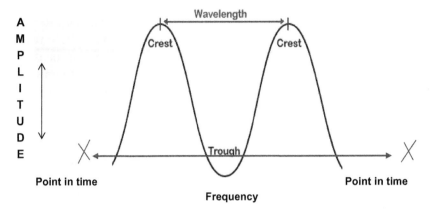

Figure 1-2 Wavelength and Frequency

the bottom of the sync pulse to the top of the white level, hence one volt peak-to-peak (p/p). And there you have it—video signals.

WHEN EVERYTHING IS BLACK AND WHITE

Two things are necessary for a camera to produce a monochrome (black-and-white) video signal: the scanning control information called synchronizing pulses and the black-and-white picture intensity information called luma. Luma is the monochrome or black-and-white portion of a video signal. This term is sometimes incorrectly called "luminance", which refers to the actual displayed brightness. Luminance ranges from pure black to pure white. Black level is the level of brightness at the darkest (black) part of a visual image—the level of brightness at which no light is emitted from a screen, resulting in pure black. Black level varies from video display to video display with better displays having a better black level. White level is the brightness of the lightest portions of an image (white areas). There are many levels of gray within the overall grayscale, ranging from slightly gray and almost white to very dark charcoal colors that are nearly black. The level of gray, white, or black in a video signal is derived from the luminance portion of the signal.

Inside the camera there are various support circuitries and an imager that converts light to an electronic signal. On the front of the camera, a lens causes light to be focused onto the imager. An easy way to grasp this may be to think of holding a magnifying glass between the sun's rays and a piece of paper. When light rays pass through the magnifying glass, the lens, they can be focused onto a specific point on the paper and start a fire. In a camera, the light travels through the lens and is focused onto the imager (minus the fire of course!). The imager converts the focused light to an electronic signal with a voltage level proportional to the brightness level of the focused image. The black-and-white portion of a video signal, which carries the information for brightness and darkness and contrast, is luminance.

The camera sends out this electronic signal similar to the way we read a book, from left to right, line after line, top to bottom, and page after page. This is called horizontal and vertical retrace. The scan lines are the portion that are visible in the image, while the retrace, or return to the start of the next line, is not. Take a moment to look at Figure 1-3, which illustrates horizontal and vertical retrace. Notice that at the end of each horizontal line, your eye retraces back to the beginning of the next line, providing the horizontal retrace. At the end of the page, your eye retraces vertically to the top of the next page, which is the vertical retrace.

The camera's support circuitry, mentioned earlier, now comes into play by adding a horizontal synchronizing (horizontal retrace) pulse at the end of each scanned line. Before each line is scanned, horizontal sync pulses set the electron beam to a locked position

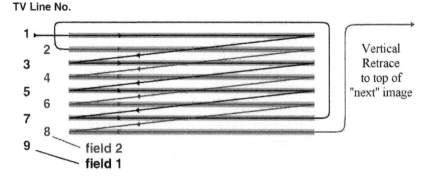

Figure 1-3 Horizontal and Vertical Retrace

so that each line of picture information starts at the same position during scanning. There is also a horizontal blanking interval, which occurs between the end of one scan line and the beginning of the next. This blanking interval is controlled by the horizontal sync pulse. When all the lines of a page have been scanned, the camera adds a vertical synchronizing (vertical retrace) pulse to the video signal and begins the next page of scanning. The vertical sync pulse controls the length of time of the vertical blanking interval. This is the period when the TV screen goes blank between the end of one field and the beginning of the second field. The combination of these two is known as composite sync.

Figure 1-4 shows the composite video signal that results from one horizontal scan line of a grayscale chart. Notice that the bars

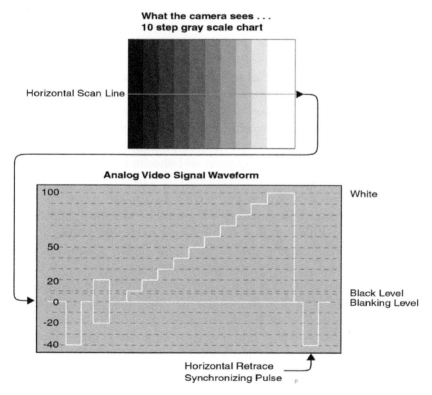

Figure 1-4   Composite Video Signal

of the grayscale chart are black on the left and white on the right, with shades of gray in the middle. Now, notice the horizontal white lines in the analog video signal waveform. You can see that each of these lines is the same width as the gray bar it represents. The white line's height above black level represents its voltage level, or how bright (what shade of gray) the bar is. The grayscale video waveform is often called a stair-step because the video signal waveform looks like a series of steps.

CREATING MOTION

Motion pictures originally set the frame rate at 16 frames per second. This was found to be unacceptable and the frame rate was increased to 24 frames per second. In Europe, this was changed to 25 frames per second, as the European power line frequency is 50 Hz.

Because video technology evolved after motion picture technology, many of the terms used in video are borrowed from the motion picture vocabulary. The concept of frames and fields is rooted in motion picture technology. For example, motion picture film is exposed at a rate of 24 images, or frames, per second. The rather low frame rate is a compromise between the amount of time needed to expose the film with enough light to make a good image and the number of frames per second necessary to provide the illusion of continuous motion. The human eye sees continuous motion, but with a very noticeable flicker in the brightness of the image. By projecting each frame twice, the flicker disappears and the human eye perceives only continuous motion.

A motion picture projector is equipped with a rotating shutter that alternately reveals and blocks the light from a bright light source. The shutter is synchronized with the mechanism that moves the film past the light source so that one frame is flashed two times onto the projection screen. See Figure 1-5. The result is that 24 frames per second are projected onto the screen two times each, or 48 fields per second.

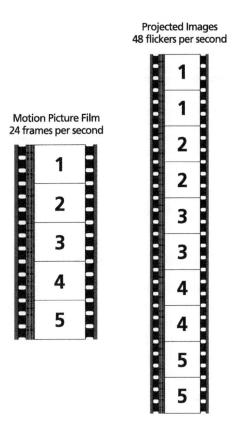

Figure 1-5   Motion Picture Projection

INTERLACE—FRAMES AND FIELDS

Like motion pictures, each frame of video is made up of two fields; therefore, there are 60 fields per second in a video stream. However, unlike motion pictures where one single frame is projected twice, each video field is generated within the camera. Two fields, field 1 and field 2, together, make one frame. Figure 1-6 illustrates how field 1 is the scan of all the odd numbered lines (1, 3, 5, 7 and so on) and field 2 is the scan of all the even numbered lines (2, 4, 6, 8 and so on). The fields are interlaced. The same process takes place in PAL cameras, except there are 50 fields, 25 frames per second.

TV Line No.

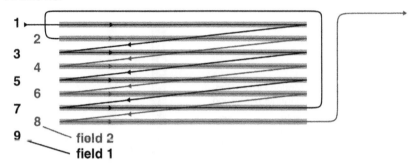

Figure 1-6   Interlaced Fields

HOW A VIDEO IMAGE IS DISPLAYED

Video is usually displayed on an analog video monitor that is comprised of a picture tube or cathode ray tube (CRT) and various support circuitries. Figure 1-7 illustrates how the composite video signal is disassembled inside the analog video monitor by a sync separator. The synchronizing pulses are converted to horizontal drive and vertical drive signals that are connected to an electromagnet.

The deflection yoke, made up of coils of wire wound around the neck of the cathode ray tube (the small end opposite the screen), generates a magnetic field and uses it to direct the electron beam in the CRT. The electromagnetic fields generated by the deflection yoke cause an electron beam inside the picture tube to reproduce the scanning pattern generated by the camera, left to right, top to bottom.

The video is applied to a control grid inside the tube to vary the intensity of the electron beam in proportion to the brightness or darkness of the original image. The more intense the beam is when it strikes the phosphor at the front of the picture tube, the brighter the phosphor glows. The less intense the beam is, the less the phosphor glows. As the electron beam scans the phosphor, left to right, top to bottom, the original image made by the camera is reproduced in the glowing phosphor, and a viewer sees a good reproduction of the camera's image.

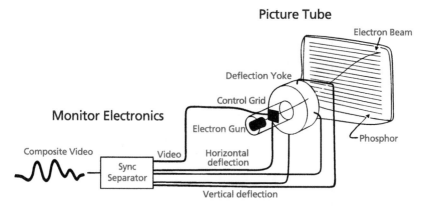

Figure 1-7   Composite Video Signal

GAMMA

Gamma is basically explained as the relationship between the brightness of a pixel as it appears on the screen and the numerical value of that pixel. Gamma correction controls the overall brightness of an image. Images that are not properly corrected can look either bleached out or too dark. Cathode-ray tubes have a peculiar relationship between the voltage applied to them and the amount of light emitted. An inverse gamma function, called gamma correction, takes place at the camera so that the video signal is non-linear for most of its journey. In other words, the transmitted signal is deliberately distorted so that, after it has been distorted again by the display device, the viewer sees the correct brightness. Figure 1-8 illustrates a video signal before and after gamma correction.

Notice the grayscale steps in the "before" video signal form a straight diagonal line as they increase in brightness voltage from left to right. The grayscale steps in the "after" video signal form a curved diagonal. The curved (non-linear) brightness steps inversely match the non-linearity of black and white picture tubes and closely match the non-linearity of color picture tubes.

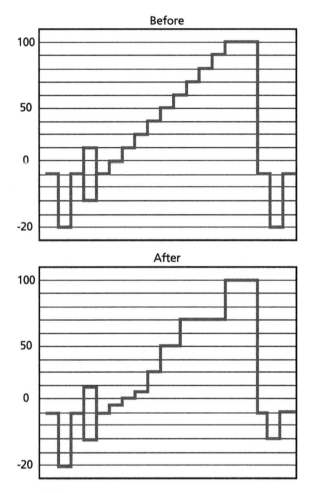

Figure 1-8   Gamma Correction

COMPONENTS OF A VIDEO SIGNAL—LUMINANCE, HUE, AND SATURATION

Analog video is an electrical signal that represents luminance, hue, saturation, and synchronizing pulses. For simplicity we have considered video as a black-and-white image up to this point. Luminance is the term that describes dark and light values in the picture we see when we view a black-and-white image. In other words, luminance is the black-and-white portion of a video signal that carries the information for brightness, darkness, and contrast. Luminance ranges from pure black to pure white. The darkest luminance level is black and the brightest luminance level is white. Most of the resolution or detail that the human eye perceives is contained in the luminance portion of an image.

When looking at a color image, two more concepts are added to the luminance of the image: hue and saturation. Hue is the term that describes the color values we see when we view a color image. We have given names to these hues or colors such as green, blue, red, purple, or yellow. Hues are actually light of differing frequencies that cause our eyes and brain to perceive different colors. Hues range from blue at the low end of the spectrum to red at the high end of the spectrum and include all of the colors we can see through green and violet and orange and yellow. Hue is the term used to state what color an object is. For example, a ripe tomato is red and the leaves on trees are green. Red and green are the hues. Saturation is a property of hue that describes how rich or intense the color is. A tomato that is just beginning to ripen is a pale red and a ripe tomato is deep red. A leaf in the spring is a light green when it first emerges and a dark green in the summer. A very intense green color is said to be rich or saturated. A very weak green color is said to be pale or pastel. A color image adds hue and saturation information to the luminance to make a complete image.

NOISE

Noise is what we call unwanted electrical signals that can be caused by the interference of electronic components in the camera and transmission lines. It can also be caused by interference from other equipment or signals that are not a part of the intended video signal. Electronic noise is present to some extent in all video signals. Broadband random noise gives the picture a snowy appearance and looks like snow or graininess over an entire image. Sources of noise include poor circuit design, excess heat, over-amplification, external influences, automatic gain control, and transmission systems. In analog and digital communications, signal-to-noise ratio, written S/N or SNR, is a measure of signal strength relative to background noise. The amount of picture information compared to the amount of noise is usually expressed in decibels (dB). Measuring SNR can be a good way of comparing the quality of video equipment.

THE ANALOG IMAGE

Once a camera has converted light into a video signal, the signal must travel outside the camera to another device such as a monitor, a VCR, or other storage device. The medium most often used for transmission is coaxial cable with a characteristic impedance of 75 Ohms (Ω). An RG-59 type coaxial cable, about $\frac{1}{4}$ inch in diameter, can carry a video signal of one volt peak-to-peak up to 1,000 feet (304.8 m) without any significant degradation of the signal. A twisted pair of wires with impedance matching transformers can carry a video signal for hundreds of feet, depending on the environment where the twisted pair wire is installed. A twisted pair of wires with active electronic amplifiers for balanced line transmission at each end can carry a video signal 3,000 feet (914 m). A fiber optic cable, similar in size or smaller than RG-59 cable, can carry a video signal several miles, depending on a variety of factors. Fiber optic cables can be used to transmit video and control signals further with no interference from common hazards such as ground loops, lightning, or man-made noise. For this reason, fiber optic cabling is often used in traffic monitoring applications.

The amount of information that can be carried in a given time period by these transmission means is called bandwidth. Bandwidth plays a very important role in the digital process, and it will be covered extensively later in the book.

One camera connected to one monitor makes up a simple system. As a camera scans each line and adds a synchronizing pulse, a monitor tracks the camera's scan by interpreting the synchronizing pulses and sprays an electron beam onto the phosphor face of the picture tube, reproducing the image. When more than

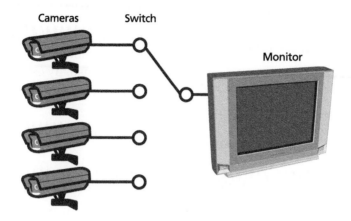

Figure 1-9 Switching

one camera needs to be displayed on a single monitor, a switch can be used to connect first one camera, then the next, and so on to the monitor. Figure 1-9 illustrates multiple cameras connected to a single monitor via a simple rotary switch.

THE IMPORTANCE OF SYNCHRONIZING

When a camera is turned on, its synchronizing (sync) generator begins to make horizontal and vertical retrace pulses, or sync pulses. As several cameras are turned on, even if they are all turned on at the same time, each camera's sync pulse generator runs to its own beat. This means that the horizontal and vertical sync pulses for each camera are occurring at different times.

As camera 1 is connected to the monitor, the monitor's deflection yoke begins to deflect the picture tube's electron beam according to the sync from camera 1. When the switch is moved to select camera 2, deflection circuits must begin to deflect the picture tube's electron beam according to the sync from camera 2. As the switch is moved to select cameras 3 and 4 in turn, the monitor's deflection yoke must again begin to deflect the picture tube's electron beam according to the new sync. When the sync timing is different for each camera, the deflection yoke has to make sudden, large adjust-

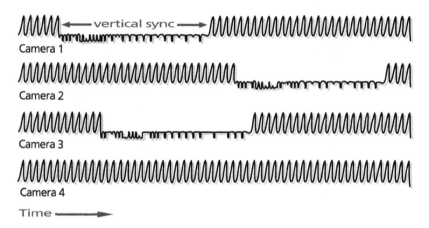

Figure 1-10   Sync

ments to track the new sync from the next camera. Figure 1-10 illustrates the video stream as it might flow from four cameras.

Notice that while the horizontal scan lines and horizontal sync pulses are fairly close to each other in time from one camera to the next, the vertical sync pulses are considerably different in time. An analog video tape recorder makes a timing signal called a control track. The function of the control track is similar to the sprocket holes in motion picture film. Control track pulses keep the tape moving from the supply reel to the take up reel at a constant speed. Control track pulses are recorded on the tape along with the video and the audio.

During playback, the control track pulses are read and compared to a reference 60-cycle signal to keep the tape motion constant. The control track signal is often generated from the vertical sync of the video being recorded. When video to tape is switched between cameras and the vertical sync pulses are not aligned in time, the control track pulse that is generated by the incoming video's vertical sync is not continuous. As a result, during playback, the picture often tears or distorts badly when the video recorder is playing between one camera and the next.

Most video cameras provide a solution for synchronization, which allows the sync pulses to line up in time either by using a synchronizing generator or by a circuit in the camera. A synchronizing generator produces horizontal and vertical synchronizing pulses that are supplied to a number of cameras. The cameras use the sync signal from the synchronizing generator to time their horizontal and vertical scans. As a result, all the cameras connected to the sync generator are reading their pictures at the same time and all the video signals arriving at the switch are synchronous.

Unfortunately, synchronizing generators can add substantial cost to a video system. The sync generator itself is a cost and each camera in a system requires at least one coaxial cable, sometimes two, from the sync generator in addition to the coaxial cable that carries the composite video back to the switcher. For this reason, sync generators are seldom used in CCTV systems. Since almost all of the cameras in a CCTV system use either primary Alternating Current (AC) power or low voltage AC power, and since both primary AC and low voltage AC have a 60-cycle alternating current, the AC itself can be used as a cheap synchronizing source.

As the AC power crosses zero on its excursion from plus to minus and back, as seen in Figure 1-11, a circuit inside the camera causes the imager to begin scanning, or reading, its next frame at the zero crossing point.

Since all the cameras in a system are connected to AC power, all of the cameras begin their scans at the same time and are subsequently synchronized vertically. This method sounds simple in theory but in practice there are some issues. Most buildings are

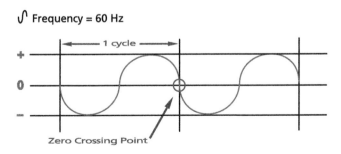

Figure 1-11   Zero Crossing Point

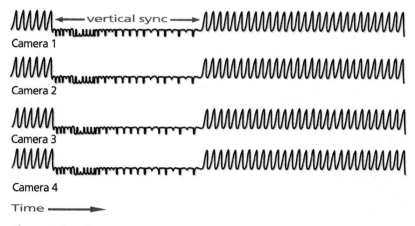

Figure 1-12   Vertical Sync

wired with 220 volt, three-phase power. Therefore, any given camera can be out of phase with another camera by 120 degrees.

In a perfect world, when video streams from synchronized cameras reach a switch or a VCR, the only thing that changes is the video itself. All the synchronizing pulses are lined up in time, and no vertical jump or roll is created when switching between cameras. Figure 1-12 illustrates video signals that are synchronized vertically.

2

What Exactly is Digital Video?

There are several technologies behind the growth of digital video for security surveillance applications. Once the personal computer gained popularity and became affordable, companies began creating security and surveillance applications for it—developing frame grabbers that convert analog video into digital images. When Ethernet and TCP/IP standards were developed, network-based applications became viable. Next, advances in wireless communications provided unprecedented mobility to surveillance applications. All of these advances in technology are possible because of video digitalization.

One of the important things to grasp about digital video is that it is simply an alternative way of carrying the same video information as an analog system. Digital video is a series or string of numbers that represent the voltage levels of an analog video signal. An ideal digital system has the same characteristics as an ideal analog system; both are completely transparent and reproduce the original applied waveform without error.

Remember the waves from chapter one? That information is now being translated into a digital language, so to speak. In fact, a very good way to understand analog and digital video technologies is to consider them as two different languages. Everyone must learn a language as a child and some people even grow up learning more than one language. We may later choose to learn more languages, which require a certain amount of time and concentration because it is not what we are used to. In the electronic industry most of us learned analog as our basic language. Now in order to understand and communicate with the digital language, we must take the time to learn it.

THE SIMPLE BREAKDOWN OF DIGITAL

The numbers used in a digital video string are called binary numbers. Binary means that there are only two possible states or conditions. It is quite simple to remember if you associate the word binary with other "bi" words such as bifocal, biplane, and bicentennial, which all refer to two of something. When referring to binary numbers, on and off or high and low are represented as one (1 = on or high) and zero (0 = off or low).

To the computer, binary digits are not really 1s and 0s. They are actually electrical impulses. Since you only have two possible switch combinations or electrical possibilities, the computer only needs various combinations of two digits to represent numbers, letters, or pixels. These two digits are visually represented by 1s and 0s. In the digital world we call these electrical impulse representations bits, or **bi**nary dig**its**. We also have bytes, which are made up of eight bits. See Figure 2-1.

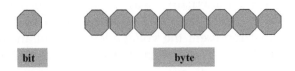

Figure 2-1   Eight Bits Make One Byte

The easiest way to understand bits is to compare them to something you are already familiar with, such as digits. A digit is a single place that can hold numerical values between 0 and 9. Digits are normally combined together in groups to create larger numbers. For example, 6357 has 4 digits. It is understood that in the number 6357 the seven is filling the "ones place", while the five is filling the "tens place", the three is filling the "hundreds place" and the six is filling the "thousands place". We all work with this type of decimal (base-10) digit system every day as a matter of course.

Computers happen to operate using the base-2 number system, known as the binary number system (just like the base-10 number system is known as the decimal number system). Where decimal digits have ten possible values ranging from 0 to 9, bits have only two possible values: 0 and 1. Therefore, a binary number is composed of only 0s and 1s like this: 1011. How do you figure out what the value of the binary number 1011 is? You do it in the same way we did it above for 6357, but using a base of two instead of a base of ten.

In the decimal counting system that we use every day there are placeholders defined by commas, which tell us how many units we are describing with a given number. See Table 2-1.

For example, the number 1000 consists of one thousands, zero hundreds, zero tens and zero units (ones). What if the placeholders had a different meaning? What if the placeholders meant this? See Table 2-2.

Table 2-1 Base of Ten

Tens of Millions	Millions	Hundreds of Thousands	Tens of Thousands	Thousands	Hundreds	Tens	Units
				1	0	0	0

Table 2-2 Base of Two

Decimal number	128	64	32	16	8	4	2	1
Binary number					1	0	0	0

Table 2-3 Binary Counting

Decimal Number	Binary Number
0	0000
1	0001
2	0010
3	0011
4	0100
5	0101
6	0110
7	0111
8	1000
9	1001
10	1010

Notice that as you read from right to left, the decimal value of the placeholder doubles. In this case, 1000 would mean eight because the column representing eight has an "on" value and the remaining numbers all have an "off" value. One (8), plus zero (4), plus zero (2), plus zero (1). Using this reasoning, the binary number 1001 would represent the number nine. One (8), plus zero (4), plus zero (2), plus one (1). 1010 would represent the number ten. One (8), plus zero (4), plus one (2), plus zero (1) or eight plus two equals ten.

You should begin to see that when using binary numbers, each bit holds the value of increasing powers of two. This makes counting in binary pretty simple. Table 2-3 provides a different view of how binary counting works. This view may make it easier to see how decimal and binary numbers are related.

When you look at the binary numbers as they increment from 0 to 10, you will see a pattern. The bit on the extreme right toggles off, on, off, on, and so on. The second bit from the right, the second bit, toggles every second increment, off, off, on, on, off, off, on, on, and so on. Because there are eight bits in a byte, we can represent 256 values ranging from 0 to 255. See Table 2-4.

For example, the numbers 00011000 represent the decimal number 24. Zero (128), plus zero (64), plus zero (32), plus **one (16),**

Table 2-4 Counting Bits

128	64	32	16	8	4	2	1
0	0	0	1	1	0	0	0

plus **one (8),** plus zero (4), plus zero (2), plus zero (1) add up to **24 or 16 + 8 = 24.** It may sound a bit confusing at first, but once you catch on it is really very simple. 256 values ranging from 0 to 255 are shown here:

$$0 = 00000000 \; 1 = 00000001 \; 2 = 00000010$$
$$254 = 11111110 \; 255 = 11111111$$

As previously mentioned, a group of eight bits is called a byte. This convention has evolved over the history of binary numbers. It is important to note that some devices provide digital performance information in bits, and some in bytes. Data speed is rarely expressed in bytes per second, and rarely is data storage or memory expressed in bits. If you are not careful, you can misunderstand the meaning of the information you are evaluating. For example, the bandwidth provided by telephone carriers is commonly expressed in multiples of bits per second (bps), and the size of a file you may want to send or receive over the phone line is expressed in multiples of bytes.

The most common convention for abbreviating bits and bytes is to use the lower case "b" for bits and the upper case "B" for bytes. A voice grade telephone line might provide a capacity or bandwidth of 64 Kbps or 64 kilobits per second, and the size of the file to be sent may be 64 KB or 64 kilobytes. 64 Kbps involves 64,000 bits, while 64 KB is describing 512,000 bits. That is a difference of 448,000 bits, which could result in a colossal misunderstanding.

Kilo or k represents 1,024 bits rounded to 1,000 for convenience. Larger amounts of bytes are described with the prefixes Mega, Giga, Terra, Peta, Exa, Zetta and Yotta, which sounds a lot like something out of a *Star Wars* movie! These become Megabyte, Gigabyte, and so on. Even shorter descriptives are derived from using singles letters as in K, M and G, written Kbytes, Mbytes, and Gbytes or KB, MB, and GB. See Table 2-5.

Table 2-5   Numeric Abbreviations

Name	Abbreviation	Size
Kilo	K	$2^{\wedge}10 = 1{,}024$
Mega	M	$2^{\wedge}20 = 1{,}048{,}576$
Giga	G	$2^{\wedge}30 = 1{,}073{,}741{,}824$
Terra	T	$2^{\wedge}40 = 1{,}099{,}511{,}627{,}776$
Peta	P	$2^{\wedge}50 = 1{,}125{,}899{,}906{,}842{,}624$
Exa	E	$2^{\wedge}60 - 1{,}152{,}921{,}504{,}606{,}846{,}976$
Zetta	Z	$2^{\wedge}70 = 1{,}180{,}591{,}620{,}717{,}411{,}303{,}424$
Yotta	Y	$2^{\wedge}80 = 1{,}208{,}925{,}819{,}614{,}629{,}174{,}706{,}176$

You can see from this chart that Kilo is about a thousand, Mega is about a million, and Giga is about a billion, and so on. So when someone says, "this computer has a 2 gig hard drive", what he/she means is "2 gigabytes", meaning approximately 2 billion bytes and exactly 2,147,483,648 bytes.

HOW DOES ANALOG VIDEO BECOME DIGITAL VIDEO?

There are a number of ways that video can be represented digitally. One way is by using Pulse Code Modulation (PCM), in which an analog waveform at the source (transmitter end) of a communications circuit is sampled (measured) at regular time intervals. In digital technology, the analog wave is sampled at some interval and then turned into numbers that are stored in the digital device. This process is called sampling. The frequency at which samples are taken is called the sampling rate or sampling frequency.

There is a general theory in engineering that you need to sample at a rate that is at least twice the fastest frequency component of the signal you are measuring. This is called the Nyquest theory. The sampling rate or number of samples per second is several times the maximum frequency of the analog waveform in hertz (Hz). In the United States, common household electrical supply is at 60 hertz. Broadcast transmission is at much higher frequency rates, usually expressed in kilohertz (KHz) or megahertz (MHz).

The result of sampling a video signal is digital video. There are many ways to accomplish sampling. The standard that has emerged for digital video sampling is the ITU-R BT.601, more commonly known as CCIR 601. ITU stands for the International Telecommunications Union, an organization established by the United Nations with members from virtually every government in the world. The ITU's mission is to set telecommunications standards and allocate frequencies for various uses around the world.

CCIR 601 is based on multiples of a fundamental 3.375 MHz sample rate. This sampling rate has been carefully chosen because of its relationship to both NTSC and PAL. Component digital video signals are sometimes referred to as 4:2:2, meaning that for every four bits that are dedicated to the Y component, two bits each are dedicated to the U & V components on both even (second 2) and odd lines (third 2) of the image. The luminance or Y channel carries most of the image detail and is, therefore, assigned more bits. The luminance signal is sampled at 13.5 MHz, four times the fundamental sampling rate. Each of the color difference signals is sampled at 6.75 MHz, two times the fundamental sampling rate. To complete the conversion, each sample is represented by a discrete number in a process known as quantizing.

A discrete unit has no part; in other words, if it is divided the result is no longer a unit. For example, there is no such thing as half a person, so people are counted in discrete numbers. Ten people can be divided only in half, fifths, and tenths, but if you try to divide them into thirds you will receive loud complaints! See Figure 2-2.

Distance is not made up of discrete units but is continuous. Consider the distance between point A and point B in Figure 2-3. Not only can we take half of the distance from A to B, we can take any part of the distance that we like; a third, a tenth, or a hundredth. See Figure 2-4.

This is true because AB is not composed of units. Every part, however small, still has a discernable length demonstrating that which is continuous is not limited by size. A discrete number, on the other hand, will always have a limit; namely, one unit. The

Figure 2-2 A Discrete Unit Has No Part

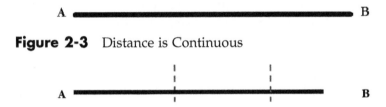

Figure 2-3 Distance is Continuous

Figure 2-4 Distance is Discrete

analog waveform is continuous and therefore must be changed into a discrete form in order to be received as digital data.

Getting back to the subject of sampling rates, examine Figure 2-5. The numbers on the left side of the scales represent voltage amplitude. To keep it simple, the values range from zero to 100 units. You also need to know that time runs from left to right. If a sine wave is one cycle of a one kHz tone, the amount of time on the grid is 1/1000 or 1 millisecond. 1 millisecond equals 1,000 microseconds (µsec.).

The small circles represent sample points. Sample points are the points at which the voltage is measured and converted into a binary number that represents the voltage level of the signal at that point. Notice that each of the samples is taken at equal time intervals. The stepped, straight-line figure in black is the digital equivalent of the sine wave reconstructed from the measurements taken at the sample points.

When you look at Figure 2-5A, you can see that the digital equivalent of the sine wave is a square wave that suggests an

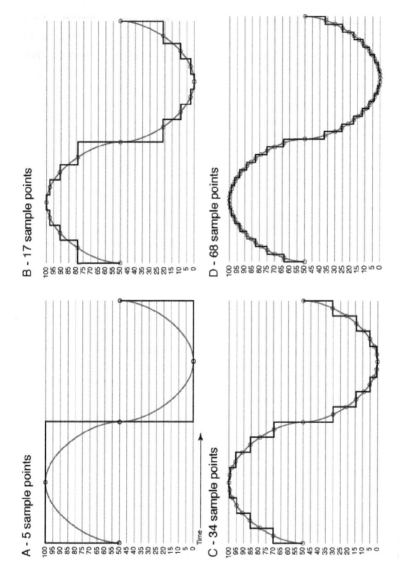

Figure 2-5 Sampling Rates

analog sine wave. The sample rate is five times the sine wave frequency, or 5,000 Hz. Increasing the sample rate to 17,000 Hz or 17 sample points as in Figure 2-5B reveals a symmetrical waveform, but the digital equivalent of the sine wave is rough. Doubling the sample rate to 34,000 Hz as in Figure 2-5C yields 34 sample points. Now the digital equivalent of the sine wave looks very much like a sine wave but it is still a little rough. Double the sample rate again to 68 Hz or 68 sample points as in Figure 2-5D and the digital equivalent looks the most like a sine wave.

Sampling could be compared with surveying the entire country for specific data. The information you receive would depend on the amount of areas sampled. In the United States, if you collect regional samples by dividing the country into the cardinal directions, you would gain four samples or four pieces of data. If you were to sample by state, you would receive fifty samples or pieces of data. Should you decide to sample by county or parish, you would gain a much larger amount of samples and consequently, you would have a very large amount of data. Obviously, it would be much easier to handle the data from the regional sampling, but how accurate are your results from the smaller sample group? On the other hand, by taking the numerous county samples, you gain a superior result at the expense of handling much larger amounts of data. Are you beginning to see the tradeoff?

It should be becoming obvious that taking more samples results in a higher resolution. The downside is that more samples also result in more data, therefore requiring more bandwidth and consequently more storage space. It is important to understand that you can convert an analog signal to representative numbers and then convert it back again to an analog signal that is an accurate representation of the original signal.

QUANTIZATION

The process of converting the sample amplitude into "bits" of data is called quantization. Quantization can occur either before or after the signal has been sampled, but usually after. Quantization is a process of approximating the continuous set of values in the image data with a finite (preferably small) set of values. It is how many levels (bits per sample) the analog signal will have to force itself into. The input to a quantizer is the original data, and the output is always one among a finite number of levels. A good quantizer ultimately represents the original signal with a minimum of loss or distortion. There are two types of quantization, scalar quantization and vector quantization. In scalar quantization, each input symbol is treated separately when producing the output. In vector quantization input symbols are organized in groups called vectors and then processed to produce output.

CREATING THE IMAGE

A computer-created image consists of many points of color called picture elements, or pixels. A pixel is the smallest element of a picture or image. Graphics monitors display pictures by dividing the display screen into thousands (or millions) of pixels, arranged in rows and columns. Pixels are so close together that they appear to be connected. To help understand this, consider the make up of a tapestry. One of the earliest forms of depicting scenes, it is actually very similar to the method of creating digital pictures. A tapestry is nothing more than fabric woven from threads of different colors that form a picture or design. The scene on a tapestry consists of vertical and horizontal threads of varying shades and sizes, which form the finished image. A digital image consists of pixels of varying shades and sizes that, when put together in a specific order, form a picture. In both cases, the detail of the image depends on the quantity of information—increasing the number of threads or pixels will produce a finer, more detailed image.

The number of bits used to represent each pixel determines how many colors or shades of gray can be displayed. For example, in 8-bit color mode, the color monitor uses 8 bits for each pixel, making it possible to display 2 to the 8th power (256) different colors or shades of gray. With 8 bits in a byte, you can represent 256 values ranging from 0 to 255, as shown here:

$$0 = 00000000 \quad 1 = 00000001 \quad 2 = 00000010$$
$$254 = 11111110 \quad 255 = 11111111$$

HOW MANY BITS DOES IT TAKE TO MAKE A BLACK-AND-WHITE PICTURE?

If there is only black-and-white information in a picture, one bit for each dot or bit per pixel (bpp) in the picture is all that is required. A dot is either black or white. The 1-bpp format is for simple black-and-white displays. 0 represents black and 1 represents white. Think of each dot in a picture as a light bulb with black representing off and white representing on. How many dots does it take to make a picture? This question leads to an understanding of resolution. Figure 2-6 illustrates a simple sketch of a face with a cook's hat on a black background.

The image here is clear and crisp. The line where the hat meets the head is smooth and straight with no jagged edges. The curves are smooth as well. This sketch was created at 300 dots per inch (dpi) resolution on a PC. Since this book is printed in a high-resolution format of 300 dpi or more, the clear, crisp lines show no jagged edges.

Figure 2-7 is the same artwork as Figure 2-6, but at a much lower resolution of 6 dpi. Without the original high-resolution image as a reference, it would be difficult to determine what this illustration is. The resolution is so low that it makes it difficult to see the mouth or the line where the chef's hat joins the head.

Figure 2-6   High Resolution

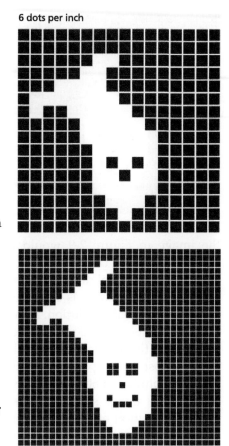

6 dots per inch

Figure 2-7 Six Dots Per Inch

Figure 2-8 Twelve Dots Per Inch

Figure 2-8 is the same image again at twice the resolution as Figure 2-7 or 12 dpi. It is a little easier to see that the picture is of a face with some sort of hat on it. The mouth is now visible but the line where the hat meets the head is still not clear. It looks as if there are ears on the head but this is just an aberration or an artifact of the low-resolution image.

Figure 2-9 is double the resolution of Figure 2-8, or 24 dpi. Now the hat looks a little like a chef's hat and the line where the hat meets the head is visible. The smile has more detail and the nose and eyes have begun to show some shape.

A resolution of 24 dpi is about 1/3 to 1/4 of the resolution of a computer graphics display monitor. Imagine how much better

Figure 2-9   Twenty-Four Dots Per Inch

the image in Figure 2-9 would become if the resolution were doubled and then doubled again.

How many dpi does it take to achieve acceptable resolution? Of course, the answer is subjective. Beauty is in the eye of the beholder and so is resolution. As a rule of thumb, 150 dpi provides "good" results. Printed images look really good at 300 dpi, which is the resolution digital photographers use when printing on photo paper. Most desktop laser printers and many ink jet printers have a resolution of 600 dpi. This is considered "excellent" quality. Professional image setter printers and premium ink jet printers have a resolution of 1200 dpi or better, but few of us can see any difference in quality between 600 dpi and 1200 dpi.

The relationship between the number of pixels in an image and the image's printed size is important to understand. Most readers can relate to the size of an image on paper, and most readers can judge the quality of what they see on paper.

The larger the pixel dimensions, the larger the number of pixels in the image and the larger the printed image size at a given resolution. Bit depth is the number of bits used to store information about each pixel. You might ask at this point, why not use the best bit depth for all images to obtain the best possible resolution? The simple answer is that the higher the bit depth, the more bits per pixel used in an image, and the more pixels used in an image, the larger the actual file is going to be. Once again, this leads to the issue of bandwidth needed to move digital information.

WHAT ABOUT GRAY?

Images made by a black-and-white video camera contain more than simple black or white dots. Black-and-white video images are comprised of black, white, and many shades of gray—as seen in Figure 2-10.

In order to convert shades of gray into numerical values for digitizing an image, we must define the shades of gray between black and white. Earlier in chapter one, Figure 1-6 portrayed a 10-step grayscale chart and the video waveform that results from one scan line of the chart by a video camera. It should be clear that

Figure 2-10   Shades of Gray

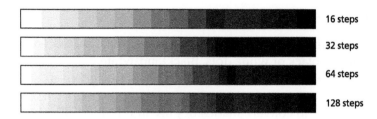

Figure 2-11   Shades of Gray in Steps

Table 2-6   Number of Bits Needed To Make Each Picture

	Fig. 3-2 6 dpi	Fig. 3-3 12 dpi	Fig. 3-4 24 dpi	Fig. 3-1 300 dpi
Dots per row	16	32	64	800
Number of rows	16	32	64	800
Total dots	256	1024	4096	639,997
Number of bits	256	1024	4096	639,997

10 shades of gray are not enough to reproduce the many shades of gray in an image like the one in Figure 2-10. How many shades are enough? Figure 2-11 illustrates the answer to this question.

64 steps of gray look almost good enough to represent continuous shading from white to black. 128 steps looks like what we need to produce continuous shading. Any more steps would probably be wasted. 256 steps are far more than what is needed to represent all of the shades of gray that the human eye can perceive.

The important piece of information about the number of shades necessary to make acceptable continuous shading is the number of bits needed to define the number of shades of gray. Table 2-6 illustrates the number of bits (and bytes) needed for 64 shade, 128 shade, and 256 shade grayscale choices.

After the digital images reach their final destination, the digital signal often has to be reverted back to an analog signal. This is achieved with a digital-to-analog converter (DAC). Basically, digital-to-analog conversion is the opposite of analog-to-digital conversion.

Table 2-7 Continuous Shading

Pixels per row	640	
Number of rows	480	
Total pixels	307,200	
# of Bits/Pixel	Total Bits	Total Bytes
64 shades—6 bits	1,843,200	230,400
128 shades—7 bits	2,150,400	268,800
256 shades—8 bits	2,457,600	307,200

When you digitize video, you decide:

1. **Frame rate—how many frames per second to capture.**
2. **Frame size—what size the frames should be.**

Even though the NTSC standard dictates 29.97 frames per second to achieve an acceptable video image, you can actually slow that rate down to 15 or 16 frames per second without diminishing the illusion of movement. The level of video resolution is also directly related to the size of the screen and the viewing distance.

Although it is possible to capture video at a size that will fill your screen—typically 640 pixels by 480 pixels—it is advisable to specify a smaller frame size in your video capture software. For example, on a monitor displaying 640 × 480 screen resolution a frame size of 160 pixels by 120 pixels would fill 1/16 of the screen. Frame sizes always maintain an aspect ratio of 4:3 to reflect the resolution of computer monitors and televisions.

A video image the size of an average 640 × 480 frame size with a resolution of 24 bits per pixel (thousands of colors) and a standard NTSC frame rate of 30 frames per second represents a little over 26 MB of data per second of video, not counting audio. This means a 1 GB hard disk could only hold about 38 seconds of video. By reducing frame size, frames per second, and bits per pixel you can make the video more manageable (i.e., make changes that would reduce the amount of information to store or transmit) at the expense of the image quality.

To calculate storage needs for digital video from black-and-white cameras, multiply the horizontal camera resolution by the vertical camera resolution. Next, divide this number by the compression factor (if compression is 10:1 use the number 10). Multiply the result by the frame-capture rate (30 frames per second multiply by 30). The final number equals the amount of bytes you will need the capacity to store per second. 720 times 480 = 345600 divided by 10 = 34560 times 30 = 1,036,800 Bytes per second (Bps) needed storage capacity per second.

Speeds at which images can be transmitted are significantly increased by digital compression, but many variables still affect the update rates. These variables include such factors as image color, movement within the image, and bandwidth of the transmission medium.

THE ELECTROMAGNETIC SPECTRUM

In order for us to see, there must be light. What we perceive as color is really a reflection from the surface of what we are looking at. The colors we perceive are actually electromagnetic radiation of various frequencies. The visible light spectrum is part of a total electromagnetic spectrum that ranges from low frequencies like radio waves in thousands of cycles per second to high frequencies like gamma rays in multi-trillions of cycles per second.

The electromagnetic spectrum is comprised of the complete range of electromagnetic frequencies from 3 kHz to beyond 300,000 THz. The electromagnetic spectrum includes, from longest wavelength to shortest: radio waves, microwaves, infrared, optical, ultraviolet, X-rays, and gamma rays. See Figure 2-12. The visible spectrum contains all the colors between infrared and ultraviolet. Infrared and ultra violet are invisible to the human eye.

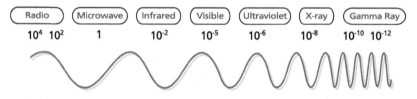

Figure 2-12   Electromagnetic Spectrum

NTSC standards have not changed significantly since their establishment, except for the addition of new strictures for color signals. NTSC signals are not directly compatible with computer systems. The NTSC standard for television defines a composite video signal with a refresh rate of 60 half-frames per second. Each frame contains 525 lines and can contain 16 million different colors. Composite video is the format of an analog television signal before it is modulated onto an RF carrier. It is a composite of three source signals called Y, U, and V (together referred to as YUV). Composite video is sometimes called CVBS for color, video, blanking, and sync, or composite video baseband signal.

When NTSC television standards were introduced the frame rate was set at 30 Hz (1/2 the 60 Hz line frequency). Then, the rate was moved to 29.97 Hz to maintain 4.5 MHz between the visual and audio carriers. Movies filmed at 24 frames per second are simply converted to 29.97 frames per second on television broadcasting.

There are three dominating video standards in use around the world, NTSC, PAL, and SECAM. These three formats have developed in different parts of the world for historic reasons. The PAL standard involves the scanning of 625 lines per frame and utilizes a wider channel bandwidth than NTSC. PAL, which stands for Phase Alternating Line, was introduced in the early 1960's and was implemented in most European countries except for France where the standard is SECAM (Sequential Couleur Avec Memoire or Sequential Color with Memory), also introduced in the 1960's. SECAM uses the same bandwidth as PAL but transmits color information sequentially. American engineers have been known to jokingly explain that SECAM stands for "System Essentially Contrary to the American Method." Generally these formats are not compatible with each other because they differ in aspects like specific scanning frequencies, number of scan lines, and color modulation techniques.

3

In the Beginning

Since the beginning of time, humanity has had an insatiable desire to communicate distances that are beyond the range of unaided hearing or sight. One of the earliest devices used in an attempt to transmit video was the Nipkow disk named for its inventor, Paul Nipkow. The Nipkow disk consisted of a circular disk with a spiral of holes cut into it. These holes were positioned so that they could scan every part of an image as it spun around. Light coming through the disk was translated to an electrical current that caused a second light to appear at the opposite end of a wire. This light passed through a second disk, which was spinning at the same speed as the original, causing a picture to be projected onto a screen. The Nipkow disk may seem very primitive now, but in its day it truly seemed like magic!

The desire to reach across distances became a reality with the development of technologies like the telegraph, telephone, and radio transmitters. As communication techniques allowed people to actually speak to each other over long distances, the wish for

Figure 3-1   Copthorne Macdonald

visual communication grew. This wish was granted in the late 1950's when an amateur radio ham developed a system of sending video signals over extensive distances that allowed him to "see" the people with whom he was speaking.

Copthorne (Cop) Macdonald (Figure 3-1) received his ham license at the age of fifteen, and his love and curiosity of the ham radio world stayed with him throughout his life. While Macdonald was working his way through the University of Kentucky's engineering school, he began looking into the feasibility of a practical SSTV (Slow Scan TV) system for ham radio operators. After addressing problems such as what sort of display tubes to use and how to get frequency response down, Macdonald gained permission from the University of Kentucky's Electrical Engineering Department to build a working model as a course project. Not only was the project a success, but his paper describing the system won first prize in the American Institute of Electrical Engineers' national competition in 1958. By developing a system of sending video signals to distant locations via radio frequency, Copthorne Macdonald charted a course in communication techniques whose benefits are still multiplying today.

In the 1960s NASA, the United States space agency, took up the challenge and improved video transmission technology for use in the space program. Though the scenes were somewhat choppy and fuzzy, the world was able to actually see early astronauts at work in their capsules as they circled the earth. The security industry has roots in the history of Slow Scan Television (SSTV) in the following lineage: A number of engineers from the NASA design team continued to research and modify remote transmission techniques and founded Robot Inc., a slow scan manufacturing company. Robot Research founded in the late 1960s in southern California and American Dynamics started in the early 1970s in upstate New York were acquired by Sensormatic Electronics Corporation, followed by Tyco International Ltd., who purchased Sensormatic in 2001.

Several well-known companies began to work on the development of video transmission products in the 60s, 70s and 80s. AT&T demonstrated its Picturephone at the World's Fair in 1964. Sony and Mitsubishi introduced monochrome versions of the videophone in the late 1980s. Unfortunately, low resolution and slow frame rates (not to mention high costs) caused these products to take a back seat to other priorities, specifically those that were more profitable.

Bringing people together from around the world and across the country for business meetings can be quite costly, both in time and money. The search for a solution to this expensive necessity guided business leaders to the advantages of teleconferencing as a communications alternative, and this growing market for remote video products stimulated further research.

Manufacturers who realized a potential market for remote video as a security surveillance product began adding their resources to the mix, along with visionaries from the world of medicine who understood the possible advantages of rapidly sending images long distances. The race was on!

THE OPTICAL ILLUSION

The process of transmitting a moving picture is dependent upon an optical illusion that makes a series of still pictures seen in rapid succession appear to be in motion. The first step in creating this illusion is to transmit one still picture or one frame of video. As discussed in chapter one, a still picture or frame of video consists of two scans containing 525 alternate horizontal lines. The camera and picture tube first scan 262.5 odd numbered lines. The picture is then scanned again to form 262.5 even numbered lines. Each half of the frame or 262.5 lines is one "field" of video. The European PAL standard involves the scanning of 625 lines per frame.

This operation of assimilating a picture, translating that picture for transmission, and then scanning that same picture at the receiving location results in the successful transmission of one full frame of video. The time involved for this process from beginning to end was initially called the "update" or "refresh" rate. Now it is commonly known as frames per second or fps.

Image updates achieved in the late 60s and early 70s were at an approximate rate of one frame every 35 seconds. Because the human brain assimilates 25 to 30 frames a second as full motion, the early technology of sending one picture each 35 seconds was far too slow to create the needed illusion.

The challenge of transmitting video evolved into a compulsion to send information at speeds that would imply full motion or "real time" to the human eye. The problem behind transmitting video information at a speed that simulates motion is in the amount of information that must be sent. This is where the value of compression comes into play. By compressing much of the information to make it smaller, it can be transmitted more efficiently.

LOSSLESS AND LOSSY COMPRESSION

Compression techniques or algorithms can be divided into two basic categories: lossless and lossy. Lossless techniques compress an image in such a manner that the final decompressed image is an exact copy of the original image. Applications such as the compression of medical X-rays or other diagnostic scans require the lossless technique to ensure that the original image is not distorted.

Lossless compression: The size of the data in its compressed form (C) in relation to the original size (O) is known as the compression ratio (R = C/O). If the inverse of the process, which is decompression, produces an exact replica of the original, the compression is known as lossless.

In this formula, you see that the lossless technique compresses information in such a manner that the final decompressed image is an exact replica of the original. This form of compression is required if you are transferring a program file or data file to be used in an application like Word or Excel as well as when transmitting information over a phone line to a facsimile machine. Initially, none of the lossless compression techniques could achieve the necessary update rates to create the illusion of full motion. For that reason, the industry turned to the lossy technique of compression for audio and video transmission, where accurate approximation is suitable.

Lossy compression: A compression scheme that intentionally loses some information in order to provide the highest compression ratio possible. Lossy compression allows only an approximation of the original to be generated. Used where the loss of information will not produce a significant degradation of information.

Unlike lossless compression, lossy techniques do not restore the original information to 100 percent of the original. The most basic lossy compression techniques simply reduce the number of bits transmitted. In lossy compression, information is analyzed and a determination is made about what loss of information will not noticeably affect the decompressed version. After concluding that the loss will not noticeably affect the image, the selected information is literally thrown away. Once this compression has been completed, it is not possible to put back the information that was removed. The missing information accounts for the differences, called artifacts, between the original and the decompressed images.

The success of video compression depends largely on the information itself. Generally some elements within the information are more common than others and most compression algorithms utilize this property, known as redundancy. The greater the redundancy the more successful the lossy compression is likely to be. Fortunately, digital video often contains a great deal of information that is redundant. Certain characteristics of the way in which humans perceive information can also be exploited to achieve higher lossy compression ratios. For example, our visual system is less sensitive to certain color information. With this eye response information, lossy compression could conceivably remove some of this color and it would not be readily perceived. The discarded information can produce minimal to drastic differences.

To obtain a better understanding of why the lossy method of compression is more suitable for video transmission, we can review the events involved in producing a moving picture. As stated, the process of creating a moving picture is dependent upon an optical illusion; specifically, a series of still pictures seen in rapid succession by the human eye appears to be in motion. Each still picture involved in the illusion is referred to as one frame of video. In order for motion to be perceived, these frames of video must pass before our eyes at a rate of approximately 25 to 30 per second. The difficulty involved in creating this optical illusion arises in transferring enough frames at a fast enough rate to perceive not just motion but fluid motion.

Here is a simplified explanation of how the speed of still frames influence the illusion of motion: Let's assume we are viewing a dog on one side of a large fence. See Figure 3-2. Presuming our frame rate is too slow to achieve the illusion of fluid motion, our next picture might show the dog on the other side of the fence. See Figure 3-2. With this information we know two things: we know the dog was originally at point A and that he

Figure 3-2   Slow Update Rate

somehow arrived at point B. Unfortunately, we don't know what events took place between these two scenes. The reason we did not see how the dog got to the other side of the fence is that the update rate is too slow. If the name of the dog is Houdini this might not be considered a problem, but in many cases, especially in a security application, this missing information could be vital.

In order to ensure that important information is not missed, we must speed up the procedure of assimilating information, process the information for transfer (i.e. compress the information to a size which can be most quickly transmitted), decompress (return the picture to its original form) the picture, and project the picture onto a display device.

The transference of video information is achieved by a variety of processes. Pulse modulation is a system of modulation involving pulses being altered and controlled to represent a message for communication. The amplitude, duration, and timing of a series of pulses are controlled in pulse-code modulation. Morse code is a very simple example of pulse-code modulation. Different modulation techniques like AM, FM, etc. represent different ways to shape or form electromagnetic radio waves. See Figure 3-3.

Amplitude Modulation—Both AM radio stations and the picture portion of a TV signal use amplitude modulation to encode information. In amplitude modulation, the amplitude of the sine wave (its peak-to-peak voltage) changes. See Figure 3-4.

Pulse Modulation

Figure 3-3   Pulse Modulation

Amplitude Modulation

Figure 3-4   Amplitude Modulation

Frequency Modulation

Figure 3-5   Frequency Modulation

Frequency Modulation—FM radio stations and hundreds of other wireless technologies (including the sound portion of a TV signal, cordless phones, and cell phones) use frequency modulation. In FM, the transmitter's sine wave frequency changes very slightly based on the information signal. See Figure 3-5.

High frequency radio waves can carry a lot of information. Very High Frequency (VHF) waves are used to carry FM radio broadcasts and Ultra High Frequency (UHF) waves are used to carry television broadcasts.

Radio Waves

A radio wave is an electromagnetic wave sometimes referred to as a Hertzian wave after Heinrich Hertz, who was the first to send and receive radio waves. James Clerk Maxwell had mathemati-

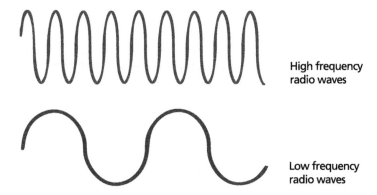

Figure 3-6   High and Low Frequency Radio Waves

cally predicted their existence in 1864, and as a professor of physics, he produced electromagnetic waves in the laboratory and measured their wavelength and velocity. Radio waves are the electromagnetic waves with the longest wavelengths and the lowest frequencies. See Figure 3-6.

Microwaves

Microwaves are very short waves of electromagnetic energy that travel at the speed of light. Microwaves are good for transmitting information from one place to another because microwave energy can penetrate haze, light rain and snow, clouds, and smoke. Radar uses microwave radiation to detect range, speed, and other characteristics of remote objects. Microwaves, used for radar, are just a few inches long. Cable TV and Internet access on coax cable as well as broadcast television use some of the lower microwave frequencies. Wireless LAN communication protocols such as IEEE 802.11 and Bluetooth also use microwaves in the 2.4 GHz ISM band, although some variants use a 5 GHz band for communication.

The Institute of Electrical and Electronics Engineers Standards Association (IEEE-SA) is the leading developer of global industry standards in a broad range of industries. The 802.11 standard

covers wireless networks. The a, b and g notations identify variations of the 802.11 standard. 802.11b was the first version to reach
the marketplace. It is the slowest and least expensive of the three.
802.11a was second to arrive, and 802.11g is a mix of both.

Infrared

Infrared radiation has a wavelength longer than visible light but
shorter than microwave radiation. Its name means "below red"
(from the Latin infra, "below") because if it were visible to the
human eye, we would see infrared directly below "red" in the
visible light portion of the spectrum. "Near infrared" light is
closest in wavelength to visible light and "far infrared" is closer
to the microwave region of the spectrum.

Infrared transmission refers to energy in the region of the
electromagnetic radiation spectrum at wavelengths longer than
those of visible light and shorter than radio waves. Infrared frequencies are higher than microwaves but lower than visible light.
Even though infrared radiation is not visible, we can feel it in the
form of heat. The longer, far infrared wavelengths are thermal.
The shorter, near infrared waves are the ones used by a TV's
remote control.

Visible Light Waves

Visible light waves make up only a small part of the electromagnetic spectrum and are the only electromagnetic waves we can
actually see with our eyes. Visible light waves appear as the colors
of the rainbow, with each color having a different wavelength.
Dispersion of visible light produces the colors red (R), orange (O),
yellow (Y), green (G), blue (B), indigo (I), and violet (V). It is
because of this that visible light is sometimes jokingly referred to
as ROY G. BIV. Red has the longest wavelength and violet has the
shortest wavelength, and if all of the waves are seen together, they
make white light.

Ultraviolet

Ultraviolet (UV) light has shorter wavelengths than visible light. Though these waves are invisible to the human eye, some insects, like bumblebees, can see them!

X-ray

As the wavelengths of light decrease, their energy increases. Because X-rays have small wavelengths we usually talk about them in terms of their energy rather than wavelength. Konrad Rontgen discovered X-rays in 1895 during an experiment with a fluorescent plate and a beam of fast electrons in a tube. He discovered that the fluorescent plate glowed even when it was a long way from the electron tube and by placing his hand in front of the fluorescent plate he created the first X-ray picture. Today, X-rays are produced in an X-ray tube by firing a beam of electrons into a Tungsten metal target.

Gamma ray

Gamma rays have the smallest wavelengths and the most energy of any other wave in the electromagnetic spectrum. These waves are generated by radioactive atoms and in nuclear explosions. Gamma rays can kill living cells and are sometimes used by physicians to kill cancerous cells.

WHAT DOES COLOR HAVE TO DO WITH IT?

Color is a complex phenomenon that has whole books dedicated to it. The objective here is to gain a very small bit of insight into what color is and what is involved in portraying color in a video situation. The significance of color is that the process of transmitting color video involves a great deal of information that will also have to be compressed in order to transmit with any useable speed. In order to transmit color video, the video signal has to provide extra color information, which consists of a synchronization signal, a luminance signal, and a chrominance signal. Monochrome images are concerned only with the first two signals that provide information about the brightness and coordination of each line that makes up the frame. These signals are called composite video because the synchronizing and luminance information are combined into a lone signal.

The color signal is called a chroma-burst and must be accurately transmitted to the receiving end with no loss of information. The chrominance portion of the signal tells a video display what color to reveal. The luminance value adjusts the color to light or dark, bright or shadowed in order to provide the correct contrast and color depth. Chrominance is abbreviated with the letter C, and we already know that luminance is abbreviated with the letter Y. A chrominance signal requires the transmission of extra data, which accounts for the larger data file related to color.

We have talked about sampling rates and pixels and how these affect the file size, but how is transmission speed affected by color? Many of the color images we see are not what they seem. What may appear to be pink, yellow, or black is actually another optical illusion. Instead of being composed of all the colors that we perceive, they are made of three primary colors (red, blue, and green) mixed together. Everything absorbs some of the light that falls on it, making it appear to be a certain color because it absorbs all of the light waves except those whose frequency corresponds to that particular color. Those waves are reflected back and cause the eye to see a particular color. The color of an object therefore depends on the frequency of the electromagnetic wave reflected.

What we perceive as color is a function of the colors contained in the source of the light that is illuminating what we see and the colors absorbed and reflected by the objects upon which the light falls. Light is focused on a sensitive part within our eyes containing two kinds of receptors: rods and cones. There are about 100 million rods and about seven million cones. Rods are sensitive to luminance or black-and-white information. Cones are sensitive to both luminance and to red, green, or blue color information.

The cones are made up of three types: one is sensitive to red-orange light, the second to green light, and the third to blue-violet light. When a single cone is stimulated, the brain perceives the corresponding color. If our green cones are stimulated, we see green; if our red-orange cones are stimulated, we see red. If both our green and red-orange cones are simultaneously stimulated, we see yellow.

The human eye cannot tell the difference between spectral yellow and some combination of red and green. Because of this physiological response, the eye can be fooled into seeing the full range of visible colors through a proportionate adjustment of just three colors: red, green, and blue. Colors are represented by bits and the more bits that are available, the more precise the color definition is portrayed. Digital video uses a non-linear variation of RGB called YCbCr. Cb represents luminance and Cr represents chrominance.

Subtractive Color

Subtractive color is the basis for printing. It is called subtractive because white, its base color, reflects all spectral wavelengths and any color added to white absorbs or "subtracts" different wavelengths. The longer wavelengths of the visible spectrum, which are normally perceived as red, are absorbed by cyan. Magenta

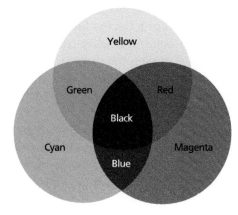

Figure   **3-7** Primary Colors in Subtractive Color System

absorbs the middle wavelengths (green), and yellow absorbs the shorter wavelengths of the visible spectrum (blue-violet). Mixing cyan, magenta, and yellow together "subtracts" all wavelengths of visible light and, as a result, we see black.

Printing inks comprised of the colors cyan, magenta, and yellow combine to absorb some of the colors from white light and reflect others. Figure 3-7 illustrates how cyan, magenta, and yellow, when printed as three overlapping circles, work to produce black as well as red, green, and blue. In practice, the black produced by combining cyan, magenta, and yellow is often not black enough to provide a large contrast range, so printers often add black ink (K) to the mix, resulting in the four color printing process sometimes known as CMYK. Cyan, magenta, and yellow are the primary colors in the subtractive color system. Red, green, and blue are the primary colors in the additive color system.

Additive Mixing

Video systems deal with subtractive color when a camera captures the light reflected from objects the same way as our eyes. But, when a video system needs to display a color image, it has to deal with a whole new way of working with color. Images that are sources of light, such as the television screen or monitor, produce color images by a process known as additive mixing. To create a color, the wavelengths of the colors are added to each other. Before any colors have been added, there is only black, which is the absence of light. On the flip side, adding all three additive primary colors in equal amounts creates white. All other colors are produced by mixing the three primary wavelengths of light in different combinations. When the three primary colors of light are mixed, intensities of the colored light are being added. This can be seen where the primary color illumination overlaps. Yellow is formed when red light added to green light is equal to the illumination of the red and green combined.

In a video signal, the color white is comprised of 30% Red, 59% Green, and 11% Blue. Since green is dominant, it is used for the luminance or black-and-white information in the picture. Note that the common symbol for luminance is the letter Y. The luminance equation is usually expressed to only 2 decimal places as $Y = 0.3R + 0.59G + 0.11B$. The letter R is of course representing red, B representing blue, and G representing green.

Instead of sending luminance (Y) and three full color signals red, green, and blue, color difference signals are made to conserve analog bandwidth. The value of green (also Y) is subtracted from the value of Red (R-Y). The value of green is also subtracted from the value of blue (B-Y). The result is a color video signal comprised of luminance Y and two color difference signals, R-Y and B-Y. Since Y (the luminance signal) is sent whole, it can be recombined with the color difference signals R-Y and B-Y to get the original red and blue signals back for display.

PICTURE QUALITY

One of the most important things to a security professional is picture quality. The effort and expense of capturing video images will be of little value if, when viewed, the image is unrecognizable. The fact is that the science of removing redundant information to reduce the amount of bits that need to be transferred would not even be necessary if we lived in a world of unlimited bandwidth. For the present, at least, this is not the case. So we must learn how to use bandwidth to its fullest advantage. Choices for high quality or high image rate result in high bandwidth requirements. Choices for lower bandwidth result in reduced image quality or reduced update rate or both. You can trade off image rate for quality within the same bandwidth.

THE BANDWIDTH WAGON

The term bandwidth is used for both analog and digital systems and means similar things but is used in very different ways. In a digital system, bandwidth is used as an alternative term to bit rate, which is the number of bits per second, usually displayed as kilobits per second. Technically, bandwidth is the amount of electromagnetic spectrum allocated to a telecommunications transmitter to send out information. Obviously, the larger the bandwidth, the more information a transmitter can send out. Consequently, bandwidth is what determines the speed and, in some cases, the clarity of the information transferred. Bandwidth is restricted by the laws of physics regardless of the media utilized. For example, there are bandwidth limitations due to the physical properties of the twisted-pair phone wires that service many homes. The bandwidth of the electromagnetic spectrum also has limits because there are only so many frequencies in the radio wave, microwave, and infrared spectrum. In order to make a wise decision about the

Figure 3-8   Traffic. Courtesy of WRI Features.

path we choose, we need to know how much information can move along the path and at what speeds. Available bandwidth is what determines how fast our compressed information can be transferred from one location to another.

Visualize digital video as water and bandwidth as a garden hose. The more water you want to flow through the hose, the bigger around the hose must be. Another example can be found in comparing driving home in rush hour traffic with transmitting video signals. If there are 500 cars, all proceeding to the same destination, how can they make the trip more expediently? A larger highway would be the obvious answer. See Figure 3-8.

If those 500 cars were traveling over four lanes as opposed to two lanes, they could move with greater speed and accuracy. Now imagine those 500 cars on an eight-lane highway. The four and eight lane highways simply represent larger bandwidths. Conversely, if there is very little traffic, a two-lane highway will be adequate. The same is true for transmitting digital data. The bandwidth requirements are dictated by the amount of data to be transferred.

4

Compression—The Simple Version

It can be difficult to get a simple answer to questions concerning the compression of video, especially when faced with making purchasing decisions. Manufacturers of video compression systems can choose from a variety of compression techniques, including proprietary technologies, and they each feel that certain attributes are more important than others. It can be easy to feel overloaded with information but at the same time feel like you are not getting any answers. The next few chapters will attempt to explain some of the common video compression idiosyncrasies so that better decisions can be made.

Compression

1. *An increase in the density of something.*
2. *The process or result of becoming smaller or pressed together.*
3. *Encoding information while reducing the bandwidth or bits required.*

Merriam-Webster Online Dictionary

Image compression is the same as data compression—the process of encoding information using fewer bits. Various software and hardware techniques are available to condense information by removing unnecessary data or what are commonly called redundancies. This reduction of information in turn reduces the transmission bandwidth requirements and storage requirements for audio, image, and full-motion video signals. The art or science of compression only works when both the sender and receiver of the information use the same encoding scheme.

The roots of compression lie with the work of the mathematician Claude Shannon, whose primary work was in the context of communication engineering. Claude Elwood Shannon is known as the founding father of the electronic communications age. Shannon investigated the mathematics of how information is sent from one location to another as well as how information is altered from one format to another. Working for Bell Telephone Laboratories on transmitting information, he uncovered the similarity between Boolean algebra and telephone switching circuits. He theorized that the fundamental unit of information is a yes-no situation in which something is or is not. Using Boolean two-value binary algebra as a code, one means "on" when the switch is closed and the power is on, and zero means "off" when the switch is open and power is off.

One of the most important features of Shannon's theory was the concept of entropy. The basic concept of entropy in information theory has to do with how much randomness is in a single or in a random event. He is also credited with the introduction of the Sampling Theory, which is concerned with representing a continuous-time signal from a (uniform) discrete set of samples. These concepts are deeply rooted in the mechanics of digital compression.

COMPRESSION IN THE 1800's

The compression of data is an idea that is not necessarily new. A compression algorithm is the mathematical process for converting data into smaller packages. An early example of a compression method is the communication system developed by Samuel Morse, known as Morse code. In 1836, Samuel Morse demonstrated the ability of a telegraph system to transmit information over wires.

The idea was to use short code words for the most commonly occurring letters and longer code words for less frequent letters. This is what is known as a variable length code. Using a variable length code, information was compressed into a series of electrical signals and transmitted to remote locations.

Morse code is a system of sending messages that uses short and long sounds combined in various ways to represent letters, numbers and other characters such as punctuation marks. A short sound is called a dit; a long sound, a dah. Written code uses dots and dashes to represent dits and dahs.

"Morse code" *World Book Online Reference Center*. 2004. World Book, Inc.

In the past, telegraph companies used American Morse Code to transmit telegrams by wire. An operator tapped out a message on a telegraph key, a switch that opened and closed an electric circuit. A receiving device at the other end of the circuit made clicking sounds and wrote dots and dashes on a paper tape. See Table 4-1. Today, the telegraph and American Morse Code are rarely used.

Compression techniques have played an important role in the evolution of telecommunication and multimedia systems from their beginnings. As mentioned in Chapter 3, pioneers of slow scan transmission of video signals have roots in the 1950s and 60s. In the 1970s, interest in video conferencing as a business tool peaked, resulting in a stimulation of research that improved picture quality and digital coding.

Early 1980s compression based on Differential Pulse Code Modulation (DPCM) was standardized under the H.120 standard. During the late 1980s, the Joint Photographic Experts Group became interested in compression of static images and they chose Discrete Cosine Transfer (DCT) as the basic unit of compression, mainly due to the possibility of progressive image transmission.

Table 4-1   Morse Code

A .-	N -.	1 .----	. .-.-.-
B -...	O ---	2 ..---	, --..--
C -.-.	P .--.	3 ...--	? ..--..
D -..	Q --.-	4 ....-	(-.--.
E .	R .-.	5	) -.--.-
F ..-.	S ...	6 -....	- -....-
G --.	T -	7 --...	" .-..-.
H	U ..-	8 ---..	_ ..--.-
I ..	V ...-	9 ----.	' .----.
J .---	W .--	0 -----	: ---...
K -.-	X -..-	/ -..-.	; -.-.-.
L .-..	Y -.--	+ .-.-.	$ ...-..-
M --	Z --..	= -...-	

This codec showed great improvement over H.120. The standard definition was completed in late 1989 and is officially called the H.261 standard.

Compression, or the process of reducing the size of data for transmission or storage, is typically achieved by the use of encoding techniques such as these just mentioned because video sequences contain a significant amount of statistical and subjective redundancy (recurring information) within frames. The ultimate goal of video compression is to reduce this information for storage and transmission by examining and discarding these redundancies and encoding a minimum amount of information. The performance of a video compression technique is significantly influenced by the amount of redundancy in the image as well as on the actual compression method used for coding.

CODECS

One second of uncompressed NTSC video requires approximately 27 MB of disk space and must definitely be compressed in order to store efficiently. Playing the video would then require decompression. Codecs were devised to handle the compression of video for storage and transmission and the decompression when it is played.

The system that compresses data is called an *encoder* or *coder*, and the decompressor is known as a *decoder*. The term codec comes from the "co" in "compressor" and the "dec" in "decompressor." When we talk about video format, we're referencing the manner in which information is stored on disks. Formats include things like AVI and QuickTime. A format does not necessarily mean anything about the video quality; it only dictates the underlying structure of a file. We'll talk more about formats in the chapter about personal computers and the Internet.

The compression of video, graphics, and audio files is accomplished by removing redundant information, thereby reducing file size. In reverse order, decompression recreates the video, graphics, and audio files. A codec is typically used when opening a video file for playback or editing, as the frames must be decompressed before they can be used. Similarly, the compressor must be used when creating a video file to reduce the size of the source video frames to keep the size of the video file to a minimum. Many codecs use both spatial and temporal compression techniques. Choosing a codec depends on the video source. For temporal compression, video that changes very little from frame to frame will compress better than video with lots of motion. With spatial compression, less detail means better compression.

Hardware codecs provide an efficient way to compress and decompress video files to make them faster and require fewer central processing unit (CPU) resources than corresponding software codecs. Using a hardware compression device can supply high-quality video images but requires viewers to have the same decompression device in order to watch it. Software codecs are less expensive, and freeware versions are often available. Viewing images compressed by software usually only require a copy of the software at the viewers end. The drawback to software codecs is that they can be CPU intensive.

Compression coder-decoders (codecs) are based upon one of four techniques for accomplishing lossy compression: (1) vector quantization, (2) fractals, (3) discrete cosine transform (DCT), and (4) wavelets. Each of these four compression techniques has advantages and disadvantages.

1. Vector quantization is a lossy compression that looks at an array of data and generalizes what it sees. Redundant data is compressed, preserving enough information to recreate the original intent.
2. Fractal compression, also a lossy compression, detects similarities within sections of an image and uses a fractal algorithm to generate the sections. Fractals and vector quantization require significant computing resources for compression but are quick at decompression.
3. DCT samples an image, analyzes the frequency components, and discards those that do not affect the image. Like DCT, discrete wavelet transform (DWT) mathematically transforms an image into frequency components. DCT is the basis of standards such as JPEG, MPEG, H.261, and H.263.
4. Wavelet mathematically transforms an entire image into frequency components that work on smaller pieces of data, resulting in a hierarchical representation of an image, where each layer represents a frequency band.

COMPRESSION SCHEMES

The principle behind compression is a simple one—convert data (using a recipe or algorithm) into a format requiring fewer bits than the original for transmission and storage. The data must be able to be returned to a good approximation of its original state. There are many popular general-purpose lossless compression techniques that can be applied to any type of data. We will examine a few here. Please do not expect to fully understand the intricacies of these techniques from the very brief explanations and examples; rather take from them the concept of various types of coding methods. In the future, when you see these or other terms relating to compression formats, you will understand the theories if not the specific complexities.

Run-length Encoding   Run-length encoding (RLE) is a simple form of data compression where strings of data occur consecutively and are stored as single data values rather than as the original string. This compression technique works by replacing consecutive incidences of a character with the character coming first and followed by the number of times the character is repeated consecutively. For example, the string 2222211111000000 is represented by 251506. The character or symbol 2 is followed by a 5 indicating the 2 appears 5 times, the 1 is followed by 5 for the 5 times it appears, and the 0 is followed by 6 for 6 times.

Clearly this compression technique is most useful where symbols appear in long runs. RLE replaces consecutive occurrences of a symbol with the symbol, followed by the number of times it is repeated. This system uses the idea that when a very long string of identical symbols appear, one can replace this long string by saying X appears 10 times. Stated another way, it replaces multiple occurrences of one value by one occurrence and the number of repetitions.

RLE takes advantage of the fact that data streams contain long strings of ones and long strings of zeros. RLE compresses the data by sending a pre-arranged code for string of ones or string of zeros followed by a number for the length of the string. The space indicated by the arrow in the following string of code represents the amount of compression achieved:

Original: 0001 0000 1111 1111 1111 1111 1111 1111 1111 1111
      1111 1111 0001 0000 0001   ←————————→
Compressed: $0 \times 3, 1, 0 \times 4, 1 \times 40, 0 \times 3, 1, 0 \times 7, 1$

This compression technique is most useful where symbols appear in long runs. RLE would not be as efficient if the symbols were not repetitious as in the following example, which shows the coded version as longer than the original version.

Original: 0 11 0 0 01 1 111 1 0 0 00 101 00   ←————————→
Compressed: $0, 1 \times 2, 0, 0, 01, 1, 1 \times 3, 1, 0, 0, 0 \times 2, 101, 0 \times 2$

Relative Encoding Relative encoding is a transmission technique that improves efficiency by transmitting the difference between each value and its predecessor, in place of the value itself. Simply put, each value is relative to the value before it. For example, if you had to compress the number string 15106433003, it would be transmitted as 1 + 4-4-1 + 6-2-1 + 0-3 + 0 + 3. In other words, the value of the number one is transmitted and the next value is conveyed by adding four to the first value of one, which equals five. The next value, which is one, is represented by subtracting four from the five we have just achieved in the previous calculation. This method of coding results in a reduction of one-third the number of bits. Differential Pulse Code Modulation (DPCM) is an example of relative encoding. By using DPCM, an analog signal is sampled and the difference between its actual value and its predicted value, which is determined from a previous sample or samples, is then converted to a digital signal.

Variable Length Codes

It is sometimes advantageous to use variable length codes (VLC), in which different symbols may be represented by different numbers of bits. For example, Morse code does not use the same number of dots and dashes for each letter of the alphabet. In particular, E, the most frequent letter, is represented by a single dot. In general, if our messages are such that some symbols appear very frequently and some very rarely, we can encode data more efficiently (using fewer bits per message) if we assign shorter codes to the frequent symbols. Consider the alternative code for the letters A through H:

$$A = 0, C = 1010, E = 1100, G = 1110, B = 100,$$
$$D = 1011, F = 1101, H = 1111.$$

Using this code, the same message as above is encoded as follows:

100010100101101100011010100100000111001111.

This string contains 42 bits and saves more than 20 percent in space compared to the three bit per character, fixed-length code shown above. An inconvenience associated with using variable length codes is that of not always knowing when you have reached the end of a symbol in a sequence of zeros and ones. In other words, how do you know when the zeros and ones have left off of representing one piece of data and begun representing another? Morse code solves this problem by using a special separator code after the sequence of dots and dashes for each letter. Another solution is to design the code in such a way that no complete code for any symbol is the beginning (or prefix) of the code for another symbol. This kind of code is called a prefix code. In the example above, the letter A is encoded by 0 and the letter B is encoded by 100, so no other symbol can have a code that begins with 0 or with 100.

In general, we can attain significant savings if we use variable length prefix codes that take advantage of the relative frequencies of the symbols in the messages to be encoded. One particular scheme for doing this is called the Huffman encoding method, which is a form of lossless compression.

Huffman Encoding

The Huffman compression algorithm is named for its inventor, David Huffman. In computer science, Huffman coding is an entropy encoding algorithm used for data compression that finds the best possible system of encoding strings based on the comparative frequency of each character. Entropy coding refers to a variety of methods that seek to compress digital data by representing frequently reoccurring patterns with minimal bits and rarely occurring patterns with many bits. Examples of entropy include run length encoding, Huffman coding, and arithmetic coding. Entropy coding, exampled by Morse code, is one of the oldest data compression techniques around. Entropy encoding assigns codes to symbols in order to match code lengths with the probabilities of the symbols appearing, resulting in the most common symbols having the shortest codes.

Huffman is a statistical data compression technique whose symbols are reassigned their original fixed length codes on decompression. Huffman's idea is very simple as long as you know the relative frequencies to put it to use. A colorful example is shown here using the word abracadabra where the letter A is the most frequently used symbol, so it is given the shortest code representation, a single 0. If you observe that the coded length of symbols is longer than the original, you are right, but remember that the goal is to reproduce the information using the shortest possible digital code.

$$A = 0$$
$$B = 100$$
$$C = 1010$$
$$D = 1011$$
$$R = 11$$
$$ABRACADABRA = 01001101010010110100110$$

A Huffman compressor computes the probability at which certain data values will occur and then assigns the shortest codes to those with the highest probability of occurring, and longer codes to the ones that don't show up as often. This method produces a variable length code where the total number of bits required to transmit data can be made considerably less than the number required if a fixed length representation is used.

Adaptive or Conditional Compression

Adaptive compression is really just what it sounds like. It dynamically adjusts the algorithm used based on the content of the data being compressed. In other words, it adapts to its environment. Adaptations of Huffman's method, known as dynamic Huffman codes or adaptive Huffman codes, were eventually developed to overcome very specific problems. Adaptive or conditional compression has achieved impressive increases in update speed by compressing and transmitting frame-to-frame motion. This system of compression sends an initial picture to a receiver and after the first full frame is sent, only those portions of the picture that have changed are compressed and transmitted.

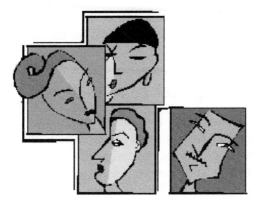

Figure 4-1 Talking Heads

A reduction in image update is the result of only small portions of the picture changing. Unfortunately, if there is a significant change to the original picture such as several persons entering the camera view, the update time will increase in direct proportion to the amount of picture changes. Conditional compression was originally very popular in the video conferencing arena where only the small movements of a person's mouth were necessary to transmit. See Figure 4-1.

UNCONDITIONAL COMPRESSION

An alternative approach commonly referred to as "unconditional" video transmission involves full frame compression. This method grabs each frame of video independently, and the entire picture is compressed and transmitted to the receiver regardless of changes within the monitored area. Because this method of compression transmits every picture in its entirety, there can be no arguments as to its integrity.

There are two criteria by which each of the compression techniques discussed here can be measured: the algorithm complexity and amount of compression achieved. When data compression is used in a data transmission application, the goal is speed. The speed of the transmission relies on number of bits sent, the time it takes the encoder to generate the coded message, and the time it takes for the decoder to recover the original data.

Intraframe or Spatial Compression

Compression is achieved by taking advantage of spatial and temporal redundancies elementary to video. In plain English, spatial compression reduces the amount of information within the space of one image by removing repetitive pieces of information. Spatial compression is used to compress the pixels of one frame by itself or one frame within a sequence to eliminate unneeded information within each frame.

Spatial redundancy takes advantage of the similarity in color values shared by bordering pixels. Spatial compression, sometimes referred to as intraframe compression, takes advantage of similarities within a video frame. Intraframe compression exploits the redundancy within the image, known as spatial redundancy. Intraframe compression techniques can be applied to individual frames of a video sequence. For example, a large area of blue sky generally does not change much from pixel to pixel.

The same number of bits is not necessary for such an area as for an area with large amounts of detail, for example if the sky was filled with multi-colored hot air balloons. Spatial compression deletes information that is common to the entire file or an entire sequence within the file. It also looks for redundant information, but instead of logging every pixel in a frame, it defines the area using coordinates.

Interframe or Temporal Compression

Some compressors employ temporal compression, which makes the assumption that frames that are next to each other look very similar. Therefore, it is used only on sequences of images. Temporal compression, sometimes referred to as frame differencing or interframe compression, compares a frame of video with the one before it and eliminates unneeded information. Temporal or interframe compression makes use of the similarities between consecutive video frames.

When it can be assumed that relatively little changes from one video frame to the next, interframe compression reduces the volume of data required to express the run of data. For example, if two consecutive frames have the same background, it does not need to be stored two times. Only the differences between the two frames need to be stored. The first frame is spatially digitized in its entirety. For the next frame, only the information that has changed is digitized. Interframe compression involves an entire sequence of video frames and the similarities between frames, known as temporal redundancy. There are several interframe compression techniques that reuse parts of frames to create new frames.

Sub-sampling can also be applied to video as an interframe compression technique, by transmitting only some of the frames. Sub-sampled digital video might, for example, contain only every second frame. Either the viewer's brain or the decoder would be required to interpolate the missing frames at the receiving end. Difference coding is a simple interframe process that only updates pixels, which have changed.

A simpler way to describe temporal compression is by understanding that it looks for information that is not necessary to the human eye. Temporal compression is accomplished by comparing images on a frame-by-frame basis for changes between frames. This compression algorithm compares the first frame with the next frame to find anything that's changed. After the initial frame, it only keeps the information that does change, allowing for the removal of a large portion of the file. It does this for each frame until it reaches the end of the file.

When there is a scene change, the new frame is tagged as the key frame and becomes the comparison image for the next frames. The comparison continues until another change occurs and the cycle begins again. The file size increases with every addition of a new key frame. This means that the fewer changes in the camera view, the smaller the data to be stored or transferred.

There are several temporal or interframe compression techniques of various degrees of complexity, most of which attempt to compress data by reusing parts of frames the receiver already has to construct new frames. Both spatial and temporal compression methods reduce the overall file size, which is of course the main goal of compression. If this does not sufficiently decrease the amount of data, one can make a larger reduction in file size by reducing colors, frame rate, and finally quality.

PREDICTIVE VS. TRANSFORM CODING

In predictive coding, information is used to predict future values. With this technique, one removes the correlation between neighboring pixels and quantizes only the difference between the value of a sample and a predicted value; then the difference is coded. Differential Pulse Code Modulation, discussed earlier, is an example of predictive coding. Transform coding transforms the image from its spatial domain representation using some well-known transform and then codes the transformed values. The image is divided into blocks and the transform is calculated for each block. After the transform is calculated, the transform coefficients are quantized and coded. The transform method affords greater data compression compared to predictive methods, in exchange for more complex computation.

Some codecs incorporate motion prediction because moving objects are reasonably predictable. The first frame in a sequence is coded in the normal way for a still image, and in subsequent frames the input is the difference between the input frame and the prediction frame. The difference frame is called the prediction error frame.

In MPEG (Motion Picture Experts Group) compression, where picture elements are processed in blocks, bits are saved by predicting how a given block of pixels will move from one frame to the next. Only the motion vector information is sent. With motion prediction, several frames of the video are being processed within the compressor at any given time, which produces a delay.

Rather than simply comparing two successive frames, this technique notices moving objects in a frame and predicts where they will be in the next frame so only the difference between the prediction and the actual location needs to be stored. For common video sequences, areas of pictures in successive frames are highly correlated. Motion prediction exploits such correlations to attain better quality and lower bandwidth requirements.

When video compression made the leap from intra-frame to inter-frame techniques, the gains were minimal. As inter-frame compression became more advanced, appreciably lower bit rates were achieved meeting memory and computational requirements, but this was still costly.

Fixed Length Codes

Let's examine this example of a fixed length code. If we use the eight symbols A, B, C, D, E, F, G, and H to create all of our messages, we could choose a code with three bits per character, for example: A = 000, C = 010, E = 100, G = 110, B = 001, D = 011, F = 101, and H = 111.

With this code, the message BACADAEAFABBAAAGAH (which contains eighteen characters) is encoded as the following string of 54 bits:

001000010000011000100000101000001001000000000110000111.

We know there will be 54 bits in the string because 18 characters times three bits equals 57. The American Standard Code for Information Interchange (ASCII) is a 7-bit code that was proposed by the American National Standards Institute (ANSI) in 1963 and finalized in 1968. The ASCII coding system contains 256 combinations of 7-bit or 8-bit binary numbers to represent every possible keystroke. Codes such as ASCII and the sample A-through-H code above are known as fixed-length codes because they represent each symbol in the message with the same number of bits: ASCII with seven bits per symbol and the A-through-H code having three bits per symbol.

COMPRESSION RATIO

Even though you will hear the term compression ratio used quite frequently in connection with digital video, you do not necessarily want to be sold or sell digital video systems based upon compression ratios. A compression ratio is simply a figure that describes the difference between information in and information out. It describes a numerical representation of the original video information compared to the compressed version of the same information.

For example, a compression ratio of 200:1 describes the original video with the numeric value of 200. In comparison, the compressed video is represented by the lower number. As more compression occurs, the numerical difference between the two numbers increases. The compression ratio is equal to the size of the original image divided by the size of the compressed image. Remember the formula (R = C/O) from chapter three? A 10 MB file that compresses to 2 MB would have a 5:1 compression ratio. Another way to look at it is that MPEG4 video compressed to a 30:1 ratio allows the storage of 30 compressed frames in the same space as a single uncompressed frame.

In most cases, the video quality decreases as the compression ratio increases. This is the obvious result of throwing away more and more information to achieve compression. Think of it in terms of making orange juice from fresh oranges. See Figure 4-2. Six

Figure 4-2   Six To One Ratio

oranges may be squeezed down to make one cup of orange juice; thus the compression ration is six to one or 6:1. First you have six whole oranges. In order to receive the benefit of the oranges (the juice) you may discard the skin, seeds, and most of the pulp. The juice that remains is still identifiable as orange, and it takes about one sixth of the space to store one glass of juice as opposed to six oranges.

In terms of video data, a high compression ratio is not necessarily a good thing. The greater the amount of compression, the more data that has been discarded, and the more the original picture degraded. The type of compression technique used can also affect the results. A video stream that is compressed using MPEG at 100:1 may look better than the same video stream compressed using JPEG at 100:1. It is of the utmost importance that a video system is considered on the merits of its performance in the actual environment where it will be used, not how well it does somewhere else.

The compression technique in use and camera placement are usually the two major influencing factors when determining the evidentiary value of a video image. When it comes to digital video for security purposes, you want sharp images from the best angle possible. Everything else is secondary.

5

More on Digital Video Compression

In a perfect world, video could be transmitted without any changes or adjustments to the data whatsoever. The reality is that storing and transmitting uncompressed raw video is not a good idea because it requires too much storage space to store and bandwidth to transmit. Therefore the need for compression exists. The actual compression of video is achieved by a collection of mathematical processes that manipulate the content of an image. There are several different processes to achieve compression; in each case, the aim is to decrease the amount of data required to represent the image in a recognizable facsimile of the original. The type of video compression used for security surveillance varies between manufacturers and by products. Some types of video compression are proprietary and not compatible with other systems.

Standards ensure interoperability and increase utility and ease of use by enabling products to work together and communicate with each other. This means that products that comply with standards, no matter who develops them, are able to work with other products that comply with the same standards. There are several organizations responsible for defining standards, including the American National Standards Institute, the International Organization for Standardization, and the International Electrotechnical Commission.

The definition of a standard by the American National Standards Institute (ANSI) is "a set of characteristics or qualities that describes the features of a product, process or service." ANSI is a private, non-profit organization that administers and coordinates the U.S. voluntary standardization and conformity assessment system. The Institute's mission is to enhance both the global competitiveness of U.S. business and the U.S. quality of life by promoting and facilitating voluntary consensus standards and conformity assessment systems and safeguarding their integrity.

The International Organization for Standardization (ISO) in Geneva is the head organization for many national standardization bodies, including:

- **DIN**—Deutsches Institut fuer Normung, Germany
- **BSI**—British Standards Institution, United Kingdom
- **AFNOR**—Association francaise de normalisation, France
- **UNI**—Ente Nazionale Italiano di Unificatione, Italy
- **NNI**—Nederlands Normalisatie-instituut, Netherlands
- **SAI**—Standards Australia International
- **SANZ**—Standards Association of New Zealand, New Zealand
- **NSF**—Norges Standardiseringsforbund, Norway
- **DS**—Dansk Standard, Denmark

The ISO is the world's largest developer of standards: a nongovernmental organization that works to promote the development of standardization to facilitate the international exchange of goods and services and spur worldwide intellectual, scientific, technological, and economic activity. The ISO is the source of ISO 9000, ISO 14000, and more than 14,000 International Standards for business, government, and society. The ISO is made up of a network of national standards institutes from 146 countries working in partnership with international organizations, governments, and industry, including business and consumer representatives.

International Electrotechnical Commission (IEC) is an international standards and assessment body for the fields of electrotechnology. Established in 1988, the IEC prepares and publishes international standards for all electrical, electronic, and related technologies. These serve as a basis for national standardization and as references when drafting international tenders and contracts.

DO YOU HAVE ALGORITHM?

Before we look at individual compression standards, let's clear up the commonly used term algorithm, which is used similarly as the term standard. An algorithm, which can be implemented either in software or hardware, refers to the process for doing a task or specific function. Standards are documented agreements between groups that set rules for the development of products and services. A standard may include a specific algorithm that is used as part of the function of the product itself.

In mathematics and computer science, algorithm usually refers to a process that solves a recurrent problem. An algorithm will have a well-defined set of rules and a specific stopping point. A computer program can, in fact, be viewed as an elaborate algorithm. The word algorithm comes from the name of the mathematician Mohammed ibn-Musa al-Khwarizmi, who was part of the royal court in Baghdad and lived from about 780 to 850.

Algorithms often have repetitive steps and may require that decisions be made to complete the desired task. Different algorithms may complete the same task with difference in time, space, effort, etc. An algorithm could be compared with a recipe in that different algorithms may complete the same task utilizing a different set of instructions concerning time, space, and effort. For example, one might have a recipe for chicken soup that calls for boning the chicken first, while another recipe may suggest boning after the chicken has been boiled. Either recipe will eventually result in chicken soup when completed.

We use algorithms every day for things like baking cookies or putting together a bicycle, by following a recipe or a set of directions. So, when a seller begins talking about algorithms pertaining to his or her digital video system, they are merely referring to the particular recipe the system follows to achieve compression and other functions. Unfortunately, an algorithm cannot complete its task if it is flawed in some way, just as the recipe will not turn out as chicken soup if pork is used in place of chicken even if all the other directions are performed perfectly.

There can be a lot of confusion surrounding the terminology involved with video compression. For example, the codec should not to be confused with the video file format used to store information after it has been encoded. Codec refers to the compression/decompression procedure used by encoding tools and players. File formats are shared by encoding tools and servers to generically store encoded streams. A video codec is an algorithm that provides a lossy compression technique for video.

Architectures allow information to be traded in a standard format. Architectures in digital video allow you to specify which codecs are used and provide the overall structure and synchronization for media delivery. Codecs are the smaller encoding components that fit within an architecture. For example, QuickTime and Windows Media are architectures; Sorenson Video and MPEG-4 are video codecs. Many different codecs are available and more are being developed every day.

COMPRESSION AND TRANSMISSION STANDARDS

In recent years, a number of video compression and transmission standards have been proposed and approved by international standardization committees like the Society of Motion Picture and Television Engineers (SMPTE) and the Institute of Electrical and Electronics Engineers (IEEE). The IEEE is the largest technical professional organization in the world (in number of members), with more than 360,000 members in 150 countries (as of 2004) They are a non-profit organization, formed in 1963 by the merger of the Institute of Radio Engineers and the American Institute of Electrical Engineers. The IEEE promotes the engineering process of creating, developing, integrating, sharing, and applying knowledge about electro and information technologies and sciences for the benefit of humanity and the profession.

MPEG

The Moving Picture Experts Groups, known collectively as MPEG, is a working group responsible for developing and maintaining digital video and audio encoding standards for compressed video. MPEG has put together and uses specific procedures for the development, adoption, testing, and review of digital multimedia standards. Having these MPEG standards as international standards insures that video encoding systems will produce files that can be opened and played with any standards-compliant decoder. The major advantage of MPEG compared to other video and audio coding formats is that MPEG files are much smaller for the same quality.

Moving objects tend to be predictable and in MPEG compression, picture elements are processed in blocks. Space is saved by predicting how a block of pixels might move from frame to frame. After analyzing the video information, only the motion vector information is sent. MPEG compression utilizes three types of frame development:

- I-frame: *(Intra-frame)*

Key frames are compressed using the JPEG standard. These are frames with no past history, typically the first frame that is encoded. The I-frame is a reference point for the following B and P frames.

- P-frame: *(Predicted frame)*

Predicted frames are constructed by comparing redundancies from preceding frames.

- B-frame: *(Bi-directional frame)*

I P B B B P B B B P B B B P B B B P B B

Figure 5-1   A Sequence of Frames

Bi-directional frames are generated by referencing redundancies in either the preceding or succeeding frames. By looking for matches in the before and after frames, the B-frame can be constructed by only encoding the difference. If no good match is found, a new I-frame may be generated. A typical sequence of frames may look like this: IBBPBBPBBPBBIBBPBBPB . . . See Figure 5-1.

The prediction technique used in MPEG video is based on motion estimation. The basic premise of motion estimation is that, in most cases, consecutive video frames will be similar except for changes induced by objects moving within the frames. The result of full MPEG compression is called Group of Pictures (GOP), which must be taken as a whole in order to decompress and display them properly. A GOP describes a sequence of coded pictures bounded by I-frames (reference frames), in which Predictive (P-frames) and Bidirectional-predictive (B-frames) techniques have been used. Once coded, a GOP does not come apart into component frames easily. Full MPEG is a challenge for CCTV applications because of the industry's need for still frame picture evaluation and frame-by-frame playback.

Motion compensation is used as part of the predictive process because when an image sequence shows moving objects their motion can be measured. The information is used to predict the content of frames later in the sequence. In the MPEG standard, each picture is divided into blocks of 16×16 pixels, called a macroblock. Each macroblock is predicted from a previous or future frame, by estimating the amount of the motion in the macro block during the frame time interval. For each block in the current frame, a matching block is found in the past frame and if suitable, its motion vector is substituted for the block during transmission. Each macroblock is divided into four blocks, which are divided into eight scan lines by eight pixels. A discrete cosine transform (DCT), which is a lossy compression algorithm that regularly samples an image to get rid of information that does not affect it, is then applied to each block. DCT samples the image at regular intervals and analyzes the frequency components present in the sample. This is the basis for standards such as JPEG, MPEG, H.261, and H.263.

For an elementary look at this process let's take another glimpse at "Houdini the dog", who we introduced in chapter three. Our first picture shows a doghouse located in a garden. The entire scene is made up of the doghouse itself, grass, a few trees in the background, and a patch of blue sky. The compression process that is the same for both JPEG and the Intra-frame portion of MPEG will include an analysis of the entire scene. After this analysis is made, it may be determined that much of the green color of the grass can be discarded with minimal effect to the outcome. The same may be true for the blue sky portion of the picture. By contrast, the trees may have too many variations of color to allow for much information to be removed without distorting the final picture.

The result of this process is called the reference frame or I-frame. Now we will assume that Houdini has entered the yard and taken a sitting position in front of his house, but nothing else has changed within the scene. The P-frame is now built with only the information concerning Houdini being compressed. The information regarding the grass, sky, and trees need not be addressed as it is identical to the preceding I-frame. If Houdini takes a nap and nothing else changes within the scene for some time, the frames between the I-frame and P-frame may be filled in by B-frames that have taken information from both.

Assume that a freak blizzard blows through the garden. The sky turns dark gray and all of the leaves are blown from the trees. The grass is covered with white snow and Houdini is driven into his house. A search will be made but as none of the previous frames contain information about this new scene, an I-frame will have to be generated, beginning the process of compression over again. As you can see, the longer the scene remains the same or at least has similar components, the amount of information needed to supply frames is minimal. As dramatic differences occur within the scene, the compression process becomes more complicated, requiring more storage space and more time for transfer of data. This is certainly an oversimplified example, but hopefully the MPEG compression process can be better understood by this imaginary representation. The obvious advantage of the MPEG standard of compression is the fact that the greatest redundancy does not reside within a single frame but within a sequence of frames.

MJPEG or Motion JPEG

MJPEG is a quasi-standard consisting of chronological JPEG frames. It does not take advantage of frame-to-frame redundancies like MPEG does. With MJPEG, each frame is separately compressed into a JPEG image. The downside to MJPEG video is that the acquired information processed on equipment from one manufacturer may not be compatible with MJPEG from another manufacturer. This is partly because some proprietary systems may use key frame and difference frames to achieve results similar to interframe compression. The addition of JPEG compression means that the compressed information retains its frame integrity, but the combinations of methods used are varied.

MPEG-1

The first digital video and audio encoding standard, MPEG-1, was adopted as an international standard in 1992. This standard was designed to store and distribute audio and motion video emphasizing video quality. MPEG-1 is a common standard used for video on the Internet. MPEG-1 relies heavily on the substantial amount of redundancy within and between frames. These are spatial, spectral, and temporal redundancies, which can be compressed without significantly changing the output.

Correlation between neighboring pixel values—Spatial redundancy, as explained earlier, removes repetitive information composed of adjoining pixels.

Correlation between different color planes or spectral bands—Spectral redundancy consists of color spectra or "brightness" due to the fact that the human eye distinguishes differences in brightness more readily than it sees differences in pure color values.

Correlation between different frames in a video sequence—Temporal redundancy is the similarity of motion between frames. If motion redundancy did not exist between frames, there would be no perception of realistic motion. Level 3 of MPEG-1 is the most popular standard for digital compression of audio, commonly known as MP3.

MPEG-2

MPEG-2, published as a standard in 1994, is a high bandwidth-encoding standard based on MPEG-1 that was designed for the compression and transmission of digital broadcast television. This is the standard used by DVD players. MPEG-2 will decode MPEG-1 bit-streams. MPEG-2 was designed for high bit rate applications, and like MPEG-1, it does not work well at low bandwidths. The main difference between MPEG-1 and MPEG-2 is the encoding of interlaced frames for broadcast TV. MPEG-1 supports only progressive frame encoding and MPEG-2 provides both progressive frame and interlaced frame encoding.

MPEG-4

MPEG-4, an open standard, was released in October 1998 and introduced the concept of Video Object Planes (VOP). The result is an extremely efficient compression that is scalable from low bit rates to very high bit rates. MPEG-4 is advanced audio and video compression that is backward compatible with MPEG-1 and 2, H.261, and H.263.

MPEG-4 is designed to deliver video over a much narrower bandwidth. It uses a fundamentally different compression technology to reduce file sizes than other MPEG standards and is more wavelet based. Object encoding provides great potential for object or visual recognition indexing, based on discrete objects within a frame, and allows access to individual objects in a video sequence. For example, if you need to track particular vehicles in series of images taken from a parking garage, a properly set up MPEG-4 based system would be a good choice.

MPEG-7

MPEG-7, also called Multimedia Content Description Interface, started officially in 1997. One of main features of MPEG-7 is to provide a framework for multimedia content that will include information on content manipulation, filtering, and personalization, as well as the integrity and security of the content. Contrary to the previous MPEG standards, which described actual content, MPEG-7 will represent information about the content allowing for faster searches of video information. Visual descriptors are used to measure similarities in images or videos. These descriptors search and filter images and videos based on features such as color, object shape, object motion, and texture. High-level descriptors are used for applications like face-recognition.

MPEG-21

MPEG-21, the newest of the standards produced by the Moving Picture Experts Group, is also called the Multimedia Framework. MPEG-21 will attempt to describe the elements needed to build an infrastructure for the delivery and consumption of multimedia content. A comprehensive standard framework for networked digital multimedia, MPEG-21 includes a Rights Expression Language (REL) and a Rights Data Dictionary. Unlike other MPEG standards that describe compression coding methods, MPEG-21 describes a standard that defines the description of content and also processes for accessing, searching, storing, and protecting the copyrights of content. The goal of MPEG-21 is to define the technology needed to support users to exchange, access, consume, trade, and otherwise manipulate digital information in an efficient, transparent, and interoperable way.

JPEG

Joint Photographic Expert Group (JPEG) is a committee of the International Standards Organization (ISO), which was established to generate standards for the compression of still images. The goal of the JPEG standard was to agree upon an efficient method of image compression for both full color and grayscale images. The ISO had begun work on this standard in April 1983 in an attempt to find methods to add photo quality graphics to the text terminals. The "Joint" in JPEG stands for refers to the merger of several groupings in an attempt to share and develop their experience. JPEG achieves much of its compression by exploiting known limitations of human sight. With JPEG compression, an image is transformed by converting the values of the image to the YUV color space (Y = luminance signal, U and V = chrominance signals), allowing access to color components of the image that are the least sensitive to the human eye. The down sampling process takes fewer samples from the chrominance portions of the images than from the luminance portions. A down sampling ratio of 4:1:1 means that for every 4 samples of luminance information, each of the chrominance values is sampled once, allowing 24 bits of information to be stored as 12 bits. Due to the physiological circumstances involving sensitivity levels of color and luminance, this reduction of information should not be noticeable to the human eye.

JPEG does not work well on non-photographic images such as cartoons or line drawings, nor does it handle compression of black-and-white (1 bit-per-pixel) images or moving pictures. It does allow for compressed image size to image quality trade-off. In other words, the operator can reduce compression to achieve better resolution at the price of a slower frame rate. Compared with commercially pervasive formats, such as GIF and TIFF, JPEG images typically require anywhere from one third to one tenth the bandwidth and storage capacity.

The Central Imagery Office has chosen JPEG as the still imagery compression standard for the United States Imagery System architecture because the wide commercial acceptance of the ISO standard, coupled with its good imagery quality, will enhance interoperability. The Intelligence Community, DOD, and other agencies have a large installed base of JPEG capable systems.

JPEG 2000

JPEG 2000, which uses wavelet technology, represents the latest series of standards from the JPEG committee. Several of the fundamental differences between the common JPEG and JPEG 2000 include the option of lossless compression in JPEG 2000, the smoothness of highly compressed JPEG 2000 images, and the additional display functionality, including zooming, offered by JPEG 2000.

Two things make JPEG 2000 desirable in bandwidth-limited applications—error resilience and rate control. Error resilience is the ability of the decoder to recover from dropped packets or noise in the bit stream during file transmission. Rate control is the ability to compress an image to a specified rate.

Four modes of operation were formulated within the JPEG standard:

- Sequential Mode
- Progressive Mode
- Hierarchical Mode
- Lossless Mode

Sequential

With sequential, each image component is encoded in a single left-to-right, top-to-bottom scan. The information is than passed to an encoder, normally DCT based. This mode is the simplest and most widely implemented. The sequential JPEG image compression standard provides relatively high compression ratios while maintaining good image quality. The downside of the JPEG technique image transmission is that it may take long time to receive and display the image.

Progressive

A refinement of the basic JPEG is broken down into several passes, which are sequentially sent to a monitor. First, highly compressed, low quality data is sent to the screen and the image quality improves as more passes are completed. The first pass takes little time to execute. In the successive passes more data is sent, gradually improving the image quality.

Several methods for producing series of partial images are supported by JPEG. In this mode, the image is scanned in sections so that users can watch the image building in segments and reject images that are not of interest as they are being delivered. The progressive JPEG mode still requires long transmission times for high-resolution, complex images, or bandwidth at the receiver's end.

Hierarchical

The image is encoded at various resolutions, allowing lower resolution versions to be decoded without decoding the higher resolution versions. The advantage here is that a trade off can be made between file size and output image quality. This capability allows the image quality to be adjusted to an acceptable condition as the application requires.

Lossless

The image is encoded in a manner that allows an exact replication to be decoded after transfer. The lossless mode of JPEG does not use DCT because it would not result in a true lossless image (an image with no losses). The lossless mode codes the difference between each pixel and the predicted value for the pixel. Lossless

JPEG with the Huffman back end is not the best choice for a true lossless compression scheme because exact replication cannot be guaranteed. Although the JPEG standard covers lossless compression it is rarely, if ever, used within the security industry for video transmission.

H.261

H.261 is the video compression portion of compression standards jointly known as ITU-T H.320. ITU comes from International Telecommunications Union. The ITU Telecommunication Standardization Sector (ITU-T) is one of three Sectors of the International Telecommunication Union. ITU-T's mission is to ensure efficient and on-time production of high quality standards (recommendations) covering all fields of telecommunications. ITU-T was created on March 1, 1993 and replaced the former International Telegraph and Telephone Consultative Committee (CCITT) whose origins go back to 1865. Both public and the private sectors cooperate within ITU-T for the development of standards that benefit telecommunication users worldwide.

The ITU-T H.320 family was developed to ensure worldwide compatibility among videoconferencing and video phone systems using ISDN telephone services. H.261 is a DCT based algorithm using both intra- and inter-frame compression designed to work at data rates that are multiples of 64 K bits per second. H.261 supports CIF and QCIF resolutions.

H.261 compression is similar to the process for MPEG compression with some differences in the sampling of color information. Color accuracy is reduced, but the results are acceptable for small images. At a rate of less than 500 K bits per second, H.261 quality is better than MPEG-1.

H.263

H.263 is also a DCT based compression scheme designed with enhancements enabling better video quality over modems. H.263 is part of the ITU H.324 suite of standards, which were designed for multimedia communication over telephone lines by modem. H.263 is approximately twice as efficient as H.261 and is supported by MPEG-4.

The H.263 standard specifies the requirements for a video encoder and decoder. It does not describe the encoder or decoder itself, but it specifies the format and content of the encoded (compressed) stream. H.263 supports five resolutions. In addition to QCIF and CIF that were supported by H.261, there are SQCIF, 4CIF, and 16CIF. SQCIF is approximately half the resolution of QCIF. 4CIF and 16CIF are 4 and 16 times the resolution of CIF, respectively.

H.263.v2/H.263+

H.263.v2, also known as H.263+, is a low-bit rate compression that was designed to take the place of H.263 by adding several annexes that substantially improve encoding efficiency.

H.264

H.264 is a high compression digital video standard written by the ITU-T Video Coding Experts Group (VCEG) together with the ISO/IEC Moving Picture Experts Group (MPEG) as the product of a collective effort known as the Joint Video Team (JVT). This standard is identical to ISO MPEG-4 part 10, also known as AVC, for Advanced Video Coding. H.264 accomplishes motion estimation by searching for a match for a block from the current frame in a previously y coded frame. MPEG-2 uses only 16 × 16-pixel motion-compensated blocks, or *macroblocks*. H.264 provides the option of motion compensating 16 × 16-, 16 × 8-, 8 × 16-, 8 × 8-, 8 × 4-, 4 × 8-, or 4 × 4-pixel blocks within each macroblock. The resulting coded picture is a P-frame.

WAVELET COMPRESSION

Wavelets have been utilized in many fields including mathematics, quantum physics, electrical engineering, and seismic geology.

Recent years have seen wavelet applications like earthquake prediction and image compression. Wavelet compression, also known as Discrete Wavelet Transform (DWT), treats an image as a signal or wave, giving it the name. Wavelet algorithms process data at different scales or resolutions—they analyze according to scale.

Basically, wavelet compression uses patterns in data to achieve compression. The image information is organized into a continuous wave that has peaks and valleys and is centered on zero. After centering the wave, the transform records the distances from zero to points along the wave (these distances are known as coefficients). An average is then used to produce a simplified version of the wave, reducing the image's resolution or detail by half. The averages are averaged again, and again, resulting in progressively simpler waves.

Images compressed using wavelets are smaller than JPEG images, meaning they are faster to transfer and download. The FBI uses wavelet compression to store and retrieve more than 25 million cards, each containing 10 fingerprint impressions. 250 terabytes of space would be required to store this data before compression. Without compression, the sorting, archiving, and searching for data in these files would be nearly impossible.

The FBI tried a Discrete Cosine Transform (DCT) initially. It did not perform well at high compression ratios because blocking effects made it impossible to follow the ridge lines in the fingerprints after reconstruction. This did not happen with Wavelet Transform because it retains the details present in data. Wavelet Transforms are used in the JPEG 2000 compression standard.

Wavelet images used in forensic applications have the advantage of being frame-based, assuring authenticity of video evidence. Images are viewed from their original compressed state, including a frame-by-frame time code. There are no standards for wavelet video compression, which means that wavelet-based images from one manufacturer's system might not be able to be decompressed properly on a wavelet-based device from another manufacturer.

Wavelets and Multi-Resolution Analysis

Multi-resolution analysis (MRA) approximates several resolution levels to obtain a *multi-scale decomposition*. In other words, as the wavelet transform takes place, it generates progressively lower resolution versions of the wave by approximating the general shape and color of the image. In addition to this, it has all the information necessary to reconstruct the wave in finer detail. The idea behind MRA is that the signal is looked at very closely at first, then slightly farther away and then farther still. See Figure 5-2.

Products of a wavelet transform can be used to enhance the delivery of an image, such as providing improved quality and more efficient compression. The wavelet transform results in simplified versions of the image along with all of the information necessary to reconstruct the original because the decomposition process produces a series of increasingly simplified versions of the image. If these are played back in reverse as the image is reconstructed and displayed, the result is a picture that literally grows in size or in detail.

There are many algorithms for video compression and more are under development each day. Work continues in developing advanced codecs for various applications with different requirements, and the power of processors keeps going up as the costs come down. There may eventually be one or two standards more readily accepted, but we are still a long way from determining what those standards will be.

Two major features to look for today for a successful deployment of new digital video equipment are software programmabil-

Figure 5-2 Example of MRA

ity and system flexibility. Programmability will allow you to download different video codec formats directly onto end products, and system flexibility will enable you to switch from one digital media standard to another or even run several simultaneously. Most importantly, when considering which compression standard or combination of standards is right for you, believe what you see with your own eyes, because quality is subjective.

Despite all of the advances, remember that video quality in any given compression system is highly dependent upon the source material. Make sure that the demonstrations you receive are comparable to the real life situations you expect in your daily operation. Take a good look at the system you are considering and make sure you actually see it do everything it claims to do.

6

Internet Transmission, Networked Video, and Storage

By now, we have all been hearing new terminologies like IP-based video, web-based video, or networked video. Then there are IP-cameras and video servers, and a host of other web-based technology. If you are feeling a bit confused, you're not alone. These terms are open to interpretation in many cases because there is no genuine consensus on what they all mean.

This brings us around to the question "How digital is it?" It is possible to have analog cameras and monitors, use coaxial cable, and have a DVR, and call that a digital surveillance system. In truth, you have a digital video recorder that is converting analog signals to digital signals. A fully digital system includes CCD cameras with signal processing that send packetized video streams via Ethernet over a cat 5 LAN cable (or wireless method) to a LAN switch and into a video server, which manages and manipulates the video signal.

Not every video system claiming to be digital is the same. In many cases, existing analog cameras are kept in place to reduce the costs. Also, digital systems do not necessarily operate through a computer but may utilize a proprietary box commonly referred to as a DVR. We will talk in length about these anomalies in upcoming chapters. For now, we can unravel some of the history, terminology, and actual application of digital surveillance, starting with the computer.

The first electronic computer weighed 30 tons, used 18,000 vacuum tubes, and drained electricity from the entire West Philadelphia area when it was turned on. The first microprocessor had the same capacity and measured just 1/8" by 1/16". A microprocessor, or CPU, is an extremely small (in comparison) but powerful high speed electronic brain etched on a semiconductor chip. The microprocessor receives incoming communications from devices like keyboards and disks for processing and contains the basic logic, storage, and arithmetic functions of a computer.

Gordon Moore was employed at Fairchild Semiconductor, prior to becoming a cofounder of Intel, when he made the prediction that the amount of data that can be stored on an electronic chip would double about every 18 months. If his theory, now called Moore's Law, were to hold true, the approximate year of 2030 will find the circuits on a microprocessor measured on an atomic scale. Next will come quantum computers that will use molecular and atomic energy to perform memory and processing tasks billions of times faster than any silicon-based computer.

Why do we care about the progression of micro-processing and computers? Electronic security and especially digital video systems have become entwined with computer software and hardware to such an extent that their futures will almost certainly evolve in tandem, not to mention the capabilities that have been added by the introduction of the Internet.

THE DIFFERENCE BETWEEN INTERNET AND INTRANET

Intranet refers to any network of interconnected computers belonging to one group. It operates almost the same as the Internet and uses the same protocols and software that are used on the Internet, but is separate from the Internet. Some businesses have special intranet web sites that can only be viewed by employees in their offices, or when connected to their Virtual Private Network (VPN). Internet, on the other hand, refers to the worldwide network of interconnected computers, all of which use a common protocol known as TCP/IP to communicate with each other.

On October 24, 1995, the Federal Networking Council (FNC) unanimously passed a resolution defining the term Internet. This definition was developed in consultation with members of the Internet and intellectual property rights communities.

RESOLUTION: The Federal Networking Council (FNC) agrees that the following language reflects our definition of the term "Internet". "Internet" refers to the global information system that—(i) is logically linked together by a globally unique address space based on the Internet Protocol (IP) or its subsequent extensions/follow-ons; (ii) is able to support communications using the Transmission Control Protocol/Internet Protocol (TCP/IP) suite or its subsequent extensions/follow-ons, and/or other IP-compatible protocols; and (iii) provides, uses or makes accessible, either publicly or privately, high level services layered on the communications and related infrastructure described herein.

The migration from analog to digital video does not mean the end for existing analog equipment. Video servers convert images from existing analog CCTV cameras into digital video for network transmission.

NETWORKING

When you connect individual PCs together allowing them to access each other's information and/or resources, you have created a network. Networks are made up of hardware, network software, connecting cables, and network interface cards connected to a network server, which is the central manager of the system.

There are several kinds of networks:

- Local area networks (LANs) consist of computers that are geographically close together.
- Wide area networks (WANs) are set up with computers farther apart that are connected by communication lines or radio waves.
- Campus area networks (CANs) refer to a network of computers within a limited geographic area, like a campus or military base.
- Metropolitan area networks (MANs) implies a data network designed for a town or city.
- Home area networks (HANs) are networks restricted to a user's home that connects digital devices.

Topology is the term used for the geometric arrangement of computer systems that are networked. Topologies are either physical due to the way workstations are connected to the network or logical—referring to the arrangement of devices on a network or their method of communicating with each other. Familiar topologies include star, ring, bus, and mesh. See Figure 6-1.

When in the star format, devices connect to a central hub via nodes that communicate by passing data through the hub. Each device has its own cable connecting it to the central point much like a phone system and a central switching station. Messages go through the central computer or network server that controls the flow of data. The administrator of a star topology can give certain nodes higher priority status than others. The central computer looks for signals from these higher priority workstations before recognizing other nodes.

Ring topology is just what it sounds like: all nodes connected to a main cable in the shape of a ring through which messages are routed in one direction only. Failure of one node on the network stops data from proceeding around the ring, making it tricky to add new workstations while the network is in operation.

A bus topology connects all devices to a central cable, which is called the bus or the backbone. Since all nodes share the bus, all messages must pass through the other workstations on the way to

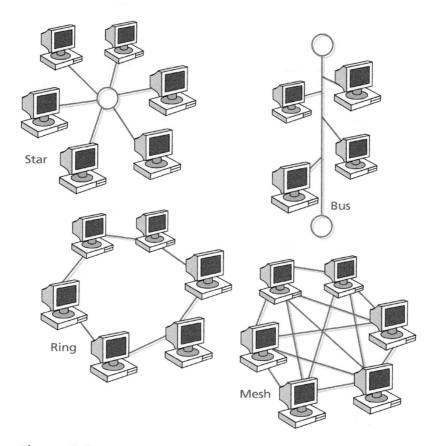

Figure 6-1 Network Designs

their destinations, but bus topologies do allow individual nodes to be out of service without disrupting service to the remaining nodes.

Mesh topology refers to devices connected with redundant interconnections between network nodes. In a true mesh topology, every node has a connection to every other node in the network. Network video technology utilizes and extends this same infrastructure for local and remote monitoring.

The network has various common parts such as hubs, switches, and routers. The hub is a common connection point that allows several pieces of equipment to share a single Ethernet connection as it forwards data packets throughout the network. A switch allows different computers or devices on a network to communicate directly with one another by filtering and forwarding packets between LAN segments. A switch can transmit data packets from different sources simultaneously. Switches can interface between 10, 100, and 1,000 Mbps. A switch may include the function of the router, determining what adjacent network point the data should be sent to. Finally, the router connects networks together and to the Internet by forwarding network packets according to their IP address.

Wireless Networks

Wireless data networks exist in a large number and variety. Some run over wireless voice networks, such as mobile telephone networks. Others run on their own physical layer networks, utilizing anything from antennas built into handheld devices to large antennas mounted on towers. A few wireless networks, like Bluetooth, are intended only to connect small devices over short distances.

NETWORKED VIDEO

When digital video systems are connected to a network, the video can be viewed by authorized parties at any other point on the network, provided they have the proper network rights and viewing software. Typically the user would enter the network address of the digital video recorder and be prompted to enter a user name and password. They could also open the user interface software, which would be preconfigured with the proper network addresses. Just as with the administrative rights of each computer and user on the network, access to the digital video can and should be restricted by user name and password, controlled by the system administrator and/or the security manager.

Within the network, Digital Video Recorders (DVRs) and Digital Video Network Servers (DVNs) perform in a similar manner to standard computer servers. Each unit will have a network address and the data on that server can be accessed through that address. DVRs and DVNs allow access to stored and live video and allow users to archive video data as needed. While they are used solely for the function of video, they are no different than any other network server in functionality and accessibility.

VIDEO DISTRIBUTION VIA THE INTERNET

Telephone lines can be used to provide the lifeline of security systems by transmitting various types of data—whether alarm information, video, or voice. Expanding telephone connectivity and increasing bandwidth have made inroads, but the real vascular system of the future is comprised of Internet connections with the PC as the heart and digital data as the blood. Advancements in digital video compression make it possible for video to be carried over typical network circuits both on the LAN and across the WAN, as well as over the Internet.

Internet Beginnings

The Defense Advanced Research Projects Agency (DARPA), which is the central research and development organization for the Department of Defense, initiated a research program pertaining to interlinking packet networks of various kinds. This was called the Internetting Project, and the resulting system of networks that emerged is what we today know as the Internet.

In 1973, development began on the protocol by a group headed by Vinton Cerf from Stanford and Bob Kahn from DARPA. The system of protocols they developed over the course of this research effort became known as the TCP/IP Protocol Suite, after two initial protocols: Transmission Control Protocol (TCP) and Internet Protocol (IP). These standards have been adopted as the standard protocols for Internetworking. Today's Internet is the result of wide area networks created in the US and the rest of the world becoming interconnected using the IP stack of protocols. TCP/IP has become the standard for networking. The open architecture allows for multiple systems to share network space and take advantage of new technology aimed at improving the capacity, reliability, scalability, and accessibility of network resources.

The resulting advantage of IP to the video surveillance world is the ability to conduct video surveillance while taking advantage of networks already in place or that will be installed for multiple purposes within a facility. Many manpower hours and dollars are saved by not having to completely rewire a facility, campus, or even a city for a new system to communicate. Video over IP is the ultimate two-for-one deal.

IP handles the actual delivery of data, while TCP is used for coordinating and organizing the data to be sent between computers via the Internet. In order for video or any other data to be transmitted via the Internet, you will need a delivery address, just as in any mail system. An IP address is a 32-bit binary number with two parts, a network address and a local address. The network address identifies the network to which the address is attached, and the local address is the specific device within that network. TCP is the tracking system that monitors individual units of data, called packets, the form in which information is efficiently routed over the Internet. TCP makes sure no packets are lost by assigning sequence numbers, which are also used to make sure that the data is delivered in the correct order.

You might compare it to sending boxes of puzzle pieces through the mail. At destination A, the puzzle is disassembled and the pieces put into various boxes. The boxes are numbered and sent on their way to destination B. Even though the boxes will travel separately, they should arrive at the same location at approximately the same time, allowing all of the pieces to be put back together to reform the original. Sending video over the Internet works in the same manner. Currently, there are two basic methods of transmitting video over the Web: streaming and downloadable video.

STREAMING VIDEO AND DOWNLOADABLE VIDEO

Streaming technologies were developed to overcome the bandwidth limitations of the Web. These are techniques for transferring

steady and continuous streams of data, making it possible to transmit and receive live video from any computer, anywhere, with
Internet access. Of course, there are protocols to be used when
transmitting data over the Internet. Hypertext Transfer Protocol
(HTTP) is the most commonly used at this time. You will recognize
the acronym from using addresses like this one: http.www.
videoonttheweb.com. The Real Time Streaming Protocol (RTSP)
handles streaming video content. Both HTTP and RTSP can be
used to deliver streaming video data, but RTSP serves as a control
protocol and as a jumping off point for negotiating transports,
such as Real-Time Protocol (RTP), over multicast or unicast
network services.

   Streaming technology sends data to the receiver continuously but does not download the entire file. This means the recipient does not have to wait to download a large file before seeing
the video because it is played as it arrives. Streaming services have
a variety of applications including online video presentations,
product demonstrations, video archive and retrieval, video broadcasting, remote monitoring, and high quality video surveillance.
The choice between streaming and downloadable video is mainly
one of duration. Since downloaded video files must be stored
somewhere (in most security applications the video will definitely
need to be stored somewhere) long files may put too much demand
on the client CPU. Downloadable video, such as QuickTime, is
stored either in memory or on disk.

   As discussed in an earlier chapter, a codec is an algorithm,
which is a recipe or list of instructions that identifies the method
used to compress data into smaller byte sizes. Codecs determine
the size of a file, the bit rate of a stream, and the way encoded
video and audio content looks and sounds. They do this by compressing an audio or video source during encoding and then
decompressing the audio or video during playback.

   When using temporal compression, where the video changes
little from frame to frame, data will compress better than video with
lots of motion. For spatial compression, the less detail in the image,
the better it can be compressed. There are many codecs available,
and many more in the pipeline. In fact, there are so many being
developed every day that it would be impossible to attempt to cover

and compare them here. We will review some popular codecs just to get a feel for what they are and what they do.

DivX Compression

A trademark of DivXNetworks, Inc., the DivX codec is a digital video compression format based on the MPEG-4 compression standard. DivX files can be downloaded over high-speed lines in a relatively short time without reducing visual quality. According to DivXNetworks, this codec is so advanced that it can reduce an MPEG-2 video (the same format used for DVD or Pay-Per-View) to ten percent of its original size and reduce video on regular VHS tapes to about one hundredth of their original size. DivX is not related to and should not be confused with DIVX (Digital Video Express), which was an attempt by Circuit City and an entertainment law firm to create an alternative to video rental in the United States.

RealVideo

As RealNetworks' cross-platform, multimedia software architecture, RealSystem is used to publish video, sound, graphics, text, music, SMIL, or Flash files over the Internet with an RTSP true streaming server. RealMedia was designed specifically for Internet multimedia. RealVideo is a streaming technology developed by RealNetworks for transmitting live video over the Internet. RealVideo works with normal IP connections and IP Multicast connections. RealVideo differs from conventional video codecs in that it has been optimized for streaming via the proprietary PNA protocol or the newer standardized Real Time Streaming Protocol. It can be used for download and play (dubbed on-demand) or for live streaming.

RealVideo was first released in 1997 as a proprietary video codec from RealNetworks. The main desktop media player for RealMedia content is the RealNetworks' RealPlayer, currently at version 10. RealNetworks claims a 30% improvement over previous generation RealVideo 9 codec with dramatically improved compression at any bit rate, on any device.

QuickTime

QuickTime is Apple's multi-platform, industry-standard, multi-media software architecture. QuickTime supports a number of compression standards including MPEG, JPEG, H.263, Microsoft Video 1, and AVI by providing software codecs that transform RAM into a compressor/decompressor for each compression type. The QuickTime architecture synchronizes video, sound, text, graphics, and music. QuickTime is a cross-platform standard, with versions running on Windows-based PCs, NT-based PCs, and UNIX-based workstations in addition to its native Apple Macintosh environment. It has an open architecture supporting many file formats and codecs, including Cinepak, Indeo, Motion JPEG, and MPEG-1. A "fast-start" feature allows users to start playing video and audio before the file is completely downloaded.

ZyGoVideo

ZyGoVideo is a QuickTime video codec especially adept at streaming video and compressing video with detailed content. ZyGoVideo works well for low bandwidth applications, especially wireless and Internet connections. ZyGoVideo allows for full screen, full frame rate video streaming using high power G5 and Pentium computers and also has a mode for streaming video to handheld and smart cell phones using processors from ARM. New product additions include handheld security solutions for commercial firms—a direct outgrowth of the Tracker and Seeker school ID security solutions.

Video for Window/Audio Video Interleaving

Video for Windows (VfW), sometimes called AVI (Audio Video Interleaving), was the first video capture and display system developed by Microsoft for the Windows operating system with a format for encoding pictures and sound for digital transmission. In this file format, blocks of audio data are woven into a stream of video frames. These files have an AVI extension.

Common Intermediate Format (CIF)

CIF is a video display standard for Internet videoconferencing that was developed by the International Telecommunication Union Telecommunication Standardization Sector (ITU-T's) as the H.261 and H.263 standards, supporting both NTSC and PAL signals. These standards are commonly used by the security industry to transmit video over the Internet.

QCIF (Quarter CIF) transfers one fourth the amount of data and is suitable for transmission systems that use telephone lines. The Q or Quarter indicates that these frames contain one quarter of the pixels as the CIF frame. This of course means that less bandwidth is needed for transmission of QCIF frames. Video quality can be broken down into three segments:

- Low quality/small image (SQCIF, QCIF)
- Medium quality/medium-sized image (CIF)
- High quality/large image (4CIF, 16CIF, SDTV)

STORING DIGITAL VIDEO

The compression of video is essential to viewing, managing, and transmitting video. The resulting reduction of data not only reduces the bandwidth needed for transmission but also reduces the space required to store it. One of the most significant factors in the growth of digital video technology in security is the ability to store larger amounts of video using less storage space and the ability to retrieve video information quickly. The terms memory and storage are often confusing, especially when describing amount. Memory refers to the amount of RAM installed in the computer, and storage refers to the capacity of the computer's hard disk.

Random Access Memory (RAM)

The computer industry commonly uses the term memory for Random Access Memory (RAM). RAM is the most common form of computer memory and is considered "random access" because you can access any memory cell directly if you know the row and column that intersect at that cell.

Read–Only Memory (ROM)

Read-only memory (ROM), also known as firmware, is an integrated circuit (IC) programmed with specific data when it is manufactured. Data stored in these chips is nonvolatile and cannot be lost when power is removed. Data stored in these chips is either unchangeable or requires a special operation to change.

Recording can be done mechanically, magnetically, or optically. Storage options are categorized as primary or secondary, volatile or non-volatile, read-only memory, Write Once, Read Many (WORM), or read-write. Each type of storage is best suited for different applications:

- Primary storage contains data actively being used.
- Secondary storage, also known as peripheral storage—the computer stores information that is not necessarily in current use.
- Volatile storage loses its contents when it loses power, non-volatile storage does not.

The good news is that the physical size of storage media keeps getting smaller while the storage capacity keeps getting larger.

Sometimes, confusion can result when referring to telecom, network, and data storage with the term Meg by itself. Mb refers to megabits and MB refers to megabytes, but the term Meg is noncommittal. Be sure to clarify. When data is transmitted in a local or wide area network, it is normally specified in megabits per second or Mbps. Data streams from digital video cameras are also specified in megabits per second. When referring to data storage, the correct term is megabytes.

HARD DISC STORAGE

In the 1950s when hard discs were first invented, they started as large disks up to twenty inches in diameter, held just a few megabytes, and were called fixed discs. The name later changed to hard disks to distinguish them from floppy disks. A hard disk drive is a machine that reads data from a hard disc and writes data onto a hard disk. They store changing digital information in a relatively permanent form. A floppy drive does the same with floppy disks, a magnetic disk drive reads magnetic disks, and an optical drive reads optical disks.

Disk drives can be either internal (within the computer) or external (outside the computer) and their performance is measured in two ways: data rate and seek time. Data rate is the number of bytes per second that the drive can deliver to the CPU, and seek time is the amount of time that passes between the time a file is requested and when the first byte of the file is sent to the CPU. Another important aspect of the capacity of the drive is the number of bytes it can hold.

In some ways, a hard disk is not that different from a cassette tape—both use the same magnetic recording techniques and both share the benefits of magnetic storage. It makes sense to use hard disks for video storage because they can store large amounts of digital data.

Techniques for guarding against hard disk failure, such as the redundant array of independent disks (RAID), were developed as precautionary measures. With this method, information is spread across several disks using techniques called disk striping (RAID Level 0), disk mirroring (RAID level 1), and disk striping with parity (RAID Level 5) to achieve redundancy and maximize the ability to recover from hard disk crashes. These are a category of disk drives that employ two or more drives in combination for fault tolerance and performance. RAID technology allows for the immediate availability of data and, depending on the RAID level, the recovery of lost data.

REMOVABLE STORAGE

Storage devices that are removable can store megabytes and even gigabytes of data on a single disk, cassette, card or cartridge. These devices fall into one of three categories: magnetic storage, optical storage, and solid state storage.

Magnetic Storage

Magnetic storage, the most common, refers to the storage of data on a magnetized medium. In most cases, removable magnetic storage consists of a mechanical device called a drive that connects to the computer. Magnetic storage is taking to paths in capacity, with some types using small cartridges with volume measured in megabytes and portable hard drives that range in the gigabytes.

Optical Storage

Basically, optical drives work by bouncing a laser beam off of the recording layer of an optical disk and registering differences in how the light is reflected back. The read-write head contains a laser, mirrors, and lenses to send and receive the reflected light. All optical drives read data in a similar way, with different technologies using different methods to record data.

Three kinds of drives record data by putting marks on a disk's recording layer: compact-disc recordable (CD-R), compact disc rewritable (CD-RW), and compact-disc read-only-memory (CD-ROM). A finished disk contains non-marked areas, indicated by marks and spaces. Each mark or space represents one bit of data. Marks are non-reflective, and spaces are reflective. Since they reflect light differently, the laser can read the recording surface and translate recorded data into 1s and 0s.

Compact Disc Recordable (CD-R) CD-R works by replacing the aluminum layer in a normal CD with an organic dye compound. This compound is normally reflective, but when the laser focuses on a spot and heats it to a certain temperature, it "burns" the dye, causing it to darken. The problem with this approach is that you can only write data to a CD-R once. After the dye has been burned in a spot, it cannot be changed back.

Compact Disc Rewritable (CD-RW) Rewritable optical drives read and write to optical disks and can be erased and rewritten many times, just like a hard disk. Some manufacturers have included Compact Disc Rewritable (CD-RW) drives on their DVR. This is NOT a good idea. With this scenario, you are allowing for the possibility that data may be altered or overwritten, destroying its evidentiary value. Some systems come with robbery buttons, which automatically download pertinent images in seconds or, in some cases, images can be range locked on the hard drive, preserving them indefinitely.

Write-Once-Read-Many   (WORM)   Write-Once-Read-Many (WORM) drives can write data to an optical disk, and then read the data over and over again. Each sector on a WORM disk can be written just once, and cannot be erased, overwritten or altered.

Solid State Storage

Solid state storage is a non-volatile, removable storage medium that employs integrated circuits rather than magnetic or optical media. It is the equivalent of large-capacity, non-volatile memory. Examples include flash memory Universal Serial Bus (USB) devices and various proprietary removable packages intended to replace external hard drives.

The main advantage of solid state storage is the fact that it contains no mechanical parts. Everything is done electronically. As a result, data transfer to and from solid state storage media takes place at a much higher speed than is possible with electromechanical disk drives. The absence of moving parts may translate into longer operating life, provided the devices are reasonably cared for and are not exposed to electrostatic discharge.

Solid state storage media lags behind electromechanical drives in terms of storage capacity. Data storage has seen a significant decrease in price with the emergence of Network Attached Storage (NAS) and SAN (Storage Area Networks), another attribute of digital technology.

Network Attached Storage (NAS)   A network attached storage (NAS) device is a server that is dedicated to file sharing. NAS does not provide any of the activities that a regular server typically provides, such as e-mail, authentication, or file management. It does allow for more hard disk storage space to be added to a network that already utilizes servers without shutting them down for maintenance and upgrades. A NAS device does not need to be located within the server but can exist anywhere in a LAN and can be made up of multiple networked NAS devices.

The Information Storage Industry Consortium (INSIC) is the research consortium for the worldwide information storage industry. INSIC membership consists of more than 65 corporations, universities, and government organizations with common interests in the field of digital information storage. Corporate membership includes major information about storage product manufacturers and companies from the storage industry infrastructure. The Storage Performance Council and the Computer Measurement Group are two members who may provide more information about storage options and advancements.

The Storage Performance Council (SPC) is a vendor-neutral standards body focused on the storage industry. It has created the first industry-standard performance benchmark targeted at the needs and concerns of the storage industry. From component-level evaluation to the measurement of complete distributed storage systems, SPC benchmarks will provide a rigorous, audited, and reliable measure of performance.

The Computer Measurement Group, commonly called CMG®, is a worldwide organization of data processing professionals committed to the measurement and management of computer systems. CMG members are primarily concerned with performance evaluation of existing systems to maximize performance and with capacity management where planned enhancements to existing systems or the design of new systems are evaluated to find the necessary resources required to provide adequate performance at a reasonable cost. Professionals charged with the measurement and management of computer systems and networks from a performance, capacity, and cost recovery standpoint may benefit from membership in CMG.

7

Guided Video
Transmission

Now that we have made the video, sampled it, and possibly compressed it, how do we get the digital video or the compressed digital video from point A to point B? And what do we need to know about the transmission medium we choose?

The quality of a data transmission is determined by the medium used for transmission and the signal itself. For guided transmission, which uses wires or cables, the medium is more important. For unguided or wireless transmission, the signal is more important. The advantage of a hardwired system is that you can usually rely on having a strong, clear signal at your destination, assuming that you are using a reasonably high quality cable and not running an unreasonable length. The disadvantage of a hardwired system is the necessity of having to run the cable between your source and destination. This can become cumbersome and costly in situations where you have to run between walls, streets, and buildings. This section deals with the three categories of guided transmission mediums available for video transmission today, which are Twisted Pair, Coaxial Cable, and Optical Fiber.

Twisted Pair

All cables, regardless of their length or quality, represent some problems when used for the transmission of video signals. The biggest problem is the need for large amounts of bandwidth. There are two main types of cable used for transmitting video signals: unbalanced (coaxial) and balanced (twisted pair). A twisted pair consists of two insulated copper wires twisted together, just as the name implies. Twisted pair includes two copper conductors, each insulated by a low-smoke, fire resistant substance. Both wires transmit and receive signal data. Because each pair carries similar electric signals, twisted pair is considered (in electrical terms) to be balanced. Balanced cable or twisted pair has been in use for many years. It is frequently used where there would be an unacceptable loss due to a long run of coaxial cable. Some of its advantages include its ability to reject unwanted interference, smaller size, and lower cost.

The twisting of wires allows each wire to have approximately the same noise level and reduces crosstalk, which is a disturbance caused by the electric or magnetic fields of one telecommunication signal affecting a signal in an adjacent circuit. When crosstalk occurs, you can hear the conversation of someone whom you did not call. You might hear half or both sides of the conversation. The occurrence of crosstalk is known as electromagnetic interference (EMI) and can be the result of things like wire placement or shielding and transmission techniques. EMI can be any electromagnetic disturbance that interrupts, blocks, or otherwise breaks down the performance of electronics or electrical equipment. Crosstalk Attenuation refers to the extent that a system resists crosstalk, bringing us back to twisted pair. It is the twisting that decreases the crosstalk interference between neighboring pairs within a cable, by using different twist lengths.

UTP and STP   There are two kinds of twisted pair used for transmission. They are unshielded (UTP) and shielded (STP) twisted pairs. Unshielded twist pair, which is subject to external electromagnetic interference, is commonly used for ordinary telephone wire. A shielded twisted pair is shielded with a metallic braid or sheath that reduces interference. Shielded twisted pair provides better performance at higher data rates for most applications, but can be more expensive and difficult to work with compared to untwisted pair. STP is not typically used for video transmission. When twisted pair cabling is used for video transmission, UTP is typically used.

Using STP cabling reduces the transmission distances, and manufacturers of the transmission and receiving components design their devices to work exclusively with unshielded cables. There are three more common types of UTP. They include Cat 3, which provides for up to 16 MHz and 16 Mbps, Cat 4 providing up to 20 MHz, and Cat 5 up to 100 MHz and 100 Mbps. More recently, Cat 5e (enhanced) and Cat 6 have been introduced and are becoming more widespread. Both provide even greater bandwidth and transmission speeds.

Cat 5

Cat 5, short for Category 5, is part of an EIA Category Specification developed to specify transmission wire used in data communications. Category 5 is specified for Ethernet 100 Base-T and 10 Base-T (100 Mbps and 10 Mbps) two conductor LANs. Cat 5 wire has a capacity of 100 Mbps and is usually terminated with an RJ 45 plug. The specification carefully controls the twist in the wire to achieve certain electrical performance levels.

Coaxial Cable

Coaxial cable was invented in 1929 and first used commercially in 1941. It is used primary by the cable television industry and is widely used for computer networks, such as Ethernet. Although more expensive than standard telephone wire, it is less susceptible to interference and can carry more data. A coax cable consists of a center conductor, a metallic outer conductor (shield), which serves as ground, an insulator covering the center conductor, and

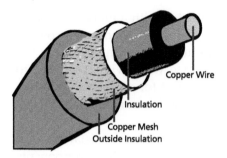

Copper Wire

Insulation

Copper Mesh
Outside Insulation

Figure 7-1 Coaxial Cable
Makeup

a plastic jacket. See Figure 7-1. It is used to transmit both analog and digital signals and has superior frequency characteristics compared to twisted pair. Its shielded concentric construction makes it less susceptible to interference and cross talk.

The commonly used terms for the two kinds of coax are Thinnet and Thicknet, sometimes called ThinWire and ThickWire. These terms refer to the larger and smaller size of coaxial cable used in Ethernet local area networks. Thicknet is 0.4 inches in diameter and has 50 ohms of electromagnetic impedance. Thinnet is 0.2 inches in diameter with the same impedance as Thicknet. Thicknet was the original Ethernet wiring, but Thinnet, which is cheaper and easier to install, is the more commonly used.

Coax is made up of two conductors that allow it to operate over a wider range of frequencies compared to twisted pair and is made in several impedances. In coaxial, impedance is measured between the inner conductor and the outer sheath. Impedance is determined by the physical construction of the cable, including the thickness of the conductors and the spacing between them. Materials used as insulators within the cable also affect impedance. Coaxial cables may be rigid with a solid sheath or flexible with a braided sheath. The inner insulator, or dielectric, has a considerable effect on the cable's properties, such as its characteristic impedance and its attenuation. The dielectric may be solid or perforated with air spaces.

The type of coaxial cable used will determine the distance that video signals can travel. These cables are categorized according to size and distance-carrying capabilities. For example, the most common coaxial cable used today is RG59/U, which carries signals for distances of approximately up to 750 or 1,000 feet, depending on which manufacturer you ask. It will be necessary to choose the correct coaxial cable for the job at hand. 75 ohm impedance cable is the standard used in CCTV systems. RG is the cable specification for use as a radio guide, and the numerical value distinguishes the specifications of each individual cable. Even though each cable has its own number, characteristics, and size, there is no difference in the way these different numbered cables work.

Coaxial cable is used primarily for the transmission of analog video. The most common place it will be seen, even with digital video systems, is from the camera to the control equipment. Since many DVRs still require an analog video input, they are still equipped with BNC connectors for the connection of coaxial cables. Some new DVRs will allow for inputs from analog and digital cameras, so until analog cameras are replaced completely with digital output cameras, coaxial cable will still be very common.

Optical Fiber

Optical fiber is a thin, flexible material used to guide optical rays. Fiber optic transmission was designed to transmit digital signals in the form of pulses of light. A fiber optic cable contains one or more hair-thin strands of glass fiber, each wrapped within a plastic tube and an external coating, and capable of transmitting messages modulated onto light waves. See Figure 7-2.

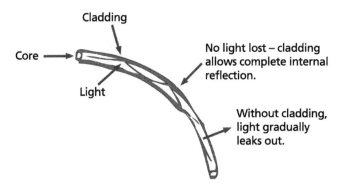

Figure 7-2 Optical Fiber With Cladding

Optical fiber is made up of a cylindrical cross-section with three concentric links:

- Core—the innermost section containing one or more very thin strands or fibers
- Cladding—a plastic or glass coating with optical properties that surrounds each strand
- Jacket—the outermost layer made of plastic and other materials that surrounds one or more claddings and protects them from environmental elements like moisture, cuts, and crushing.

Fiber optic cables are much thinner and lighter than metal wires and can carry far more information than copper wire. For longer distance, fiber optic cables can be used to transmit video signals without interference from ground loops, lightning hazards, and man made noise. Optical fibers are not affected by interference from electromagnetic sources.

FDDI   Fiber Distributed Data Interface (FDDI) refers to a set of ANSI protocols for sending digital data over fiber optic cable. These networks can support data rates of up to 100 Mbps and are often used as the basis for wide-area networks.

FDDI-2   FDDI-2 supports the transmission of voice, video, and data.

Another variation of FDDI called FDDI Full Duplex Technology (FFDT) can potentially support data rates up to 200 Mbps.

FTTP   Fiber to the Premises (FTTP) is a set of standards defining common technical requirements for extending fiber optic cabling and equipment to homes and businesses, which was begun in the U.S. in 2004 by the RBOCs. These industry standards facilitate the deployment of broadband services such as voice, video, and high-speed Internet to homes and businesses.

SONET   Synchronous Optical Network (SONET) is a standard for optical telecommunications transport. The standard defines a ranking of interface rates that allow data streams at different rates to be transmitted over a single line or media.

With a networked security installation, video can normally be viewed from any point on the network locally, as well as remotely from around the world. Access to the video information is controlled through user names and passwords, rather than restricting physical access to a monitor and/or operator keyboard. As long as you can connect to the network, there's an excellent possibility to view and manage the information coming from the cameras.

TRANSMISSION AND CONNECTION TYPES

When the transmission of digital video first became feasible, Public Switched Telephone Service (PSTN) was restricted to twisted pair telephone links with a capability of around 30 Hz to 3.4 KHz bandwidth—adequate for voice transmission but very limiting when trying to transmit video signals. The broad accessibility of PSTN service was certainly an advantage, but unfortunately, PSTN has limited bandwidth due to its employment of analog signals. These must be converted from digital and back again, greatly slowing the transmission process. The differences between analog and digital information could be compared to two different spoken languages. If you only speak Latin and you must communicate with someone who only speaks Greek you will need a translator; hence the modem.

A modem is the modulator/demodulator that translates digital signals so that they can be sent over analog lines. The receiver will also require a modem in order to translate the analog information back to a digital format. It is fairly easy to comprehend why the necessity for a translator will slow down the communication process. The speed of the translator (modem) will also be a factor. Today, modems not only run faster, they contain features like error control and data compression. Modems can also monitor and regulate information flow. These modems select the appropriate speed according to the current line conditions.

Much like the multitude of highways available for travel by car, many paths for video transmission now exist, including various modes of phone line transmission. Similar to taking a car from point A to point B, you will have decisions to make about the best route to take. Should it be the fastest, the straightest, the one with the most restaurants, or the least stop lights? There are many things to take into consideration when choosing a transmission medium for your video data as well.

- Availability
- Cost
- Speed
- Number of Users
- Reason for Connection

Costs can vary widely depending on the location, and some options are not available in all areas. In large cities, there may be a variety of options available in varying price ranges, whereas rural areas may only have a few expensive options. Certain options may only offer a specific speed while others have additional speed abilities, for an additional fee. An office with two employees may only need an ISDN connection or an ADSL connection, whereas an office with fifty employees may require a T1, Frame Relay, or ATM connection to handle the load. Of course, the reason for connection has a significant bearing on the choice of service.

The subject of data transmission involves a tremendous amount of acronyms, which will be identified throughout the chapter. Let's start with the most basic and most familiar of our choices for video transmission, the phone line. Composite video cannot be transmitted down a telephone line. It must be converted to a digital signal via a modem. The modem converts the digital signal to a series of tones, which pass down the phone lines.

POTS

The standard telephone service that most homes use is affectionately called POTS, for Plain Old Telephone Service. POTS is a standard, single line telephone service with access to the Public Switched Telephone Network (PSTN). In order to use a conventional modem for transmitting digital data and a telephone for transmitting voice at the same time on a POTS system, two lines are needed.

PSTN

Public Switched Telephone Network usually refers to the international telephone system, which is based on copper wires carrying analog voice data. It is a normal voice grade telephone line with a slow transmission speed but worldwide availability. Modern networks are not connected, and rout packets contain digitized audio voice information as it is produced.

In some countries, there is only one telephone company. In countries with many competitive phone services like the United States, telephone company refers to the entire interconnected network of phone companies. Regular modems are needed to send digital data over a PSTN line. The speed of transmission is restricted by the bandwidth of the PSTN, and the maximum amount of data that you can receive using ordinary modems is about 56 Kbps.

ISDN

The abbreviation for Integrated Services Digital Network is ISDN. This is an international communications standard for sending voice, video, and data over digital telephone lines or normal telephone wires. ISDN protocols are used worldwide for connections to public ISDN networks or to attach ISDN devices to ISDN-capable PBX systems. ISDN is a telephone company service that is supported by the ITU H.320 suite of standards and supports data transfer rates of 64 Kbps.

ISDN builds on groups of standard transmission channels. Bearer channels or B channels transmit information at comparatively high speeds. Separate Data channels or D channels carry the set-up, signaling and other user data. B channels are clear-channel pipes and D channels are packet-switched links. Packets are routed to their destination through the most expedient path. Packet switching is a communications standard in which messages or fragments of messages are individually routed between nodes, with no previously determined communication route. Each packet is transmitted individually and can follow different routes to its destination. Once all the packets forming a message arrive at the destination, they are assembled into the original message.

There are two versions of ISDN, Basic Rate Interface (BRI) and Primary Rate Interface (PRI).

- Basic Rate Interface is made up of two 64-Kbps B-channels and one D-channel for transmitting control information. Each of the two B channels is treated independently by the network, permitting simultaneous voice and data or data only connections. With specialized hardware and software, multiple B channel connections can be combined to attain rates of several Megabytes of data per minute or more. This version is referred to as ISDN 2 in Britain and ISDN 2e in Europe.
- Primary Rate Interface consists of 23 B-channels and one D-channel in the United States or 30 B-channels and one D-channel in Europe, which was designed for larger organizations. PRI service is generally transmitted through a T-1 line or an E1 line in Europe. It is also possible to support multiple PRI lines with one 64 kb/s D channel using Non-Facility Associated Signaling (NFAS), a special case of ISDN signaling in which two or more T1 PRI lines use the same D channel, and you can add a backup D channel. The NFAS option extends D-channel control to B-channels not resident on the same interface.

B-channel is the main data channel in an ISDN connection. D-channel is the ISDN channel that carries control and signal information. ISDN originally used baseband transmission but another version known as B-ISDN uses broadband transmission and is able to support transmission rates of 1.5 Mbps. This version requires fiber optics and is not readily available in many locations.

DSL

Digital Subscriber Line (DSL) is a general term for any local network loop that is digital in nature. It is a very high-speed connection that uses the same wires as a regular telephone line. The copper wires that make up regular phone lines have plenty of room to transmit more than just voice conversations. DSL takes advantage of this extra bandwidth without disturbing the voice conversations. To interconnect multiple DSL users to a high-speed network, the telephone company uses a Digital Subscriber Line Access Multiplexer (DSLAM). At the other end of each transmission, a DSLAM demultiplexes signals and sends them to individual DSL connections. On average with DSL, data is downloaded at rates up to 1.544 Mbps and you can send data at 128 Kbps. DSL service requires a special modem and a network card in your computer.

Current information on DSL technology advancements can be obtained from the DSL Forum, which is an international industry consortium of approximately 200 service providers, equipment manufacturers, and other interested parties who are focused on developing the full potential of broadband DSL. DSL Forum's Web site dedicated to providing information to end-users can be found at www.dsllife.com

ADSL   Most homes and small business users are connected to an Asymmetric Digital Subscriber Line (ADSL) that was designed to transmit digital information at a high bandwidth over existing phone lines. ADSL is different from regular phone service in that it provides a continuous or always on connection. It is called asymmetric because of the way it divides up a channel, on the assumption that most Internet users download or receive much more information than they upload or send. ADSL uses most of the available channel for receiving. ADSL supports data rates of from 1.5 to 9 Mbps when receiving data and from 16 to 640 Kbps when sending data.

An ADSL circuit connects an ADSL modem on each end of a twisted-pair telephone line, creating three information channels.

These include a high speed downstream channel, a medium speed duplex channel, and a POTS channel. The POTS channel is split off from the digital modems by filters to guarantee uninterrupted phone service. Most DSL technologies call for a signal splitter requiring a phone company visit, although it is possible to manage the splitting remotely. This is known as splitterless DSL, DSL Lite, G.Lite, or Universal ADSL and has recently been made a standard. Various forms of DSL are allowing phone companies to compete with cable modem services.

xDSL   Usually referred to as simply DSL, it is sometimes called xDSL with the "x" denoting "any" for DSL variations. It is a generic term for DSL services in that the x can be replaced with any of the letters that represent the various types of DSL technology such as these:

- Very high bit-rate DSL (VDSL), a fast connection that only works over short distances.
- Symmetric DSL (SDSL), often used by small businesses, does not allow dual use. Incoming and outgoing data-rates are the same.
- Rate-adaptive DSL (RADSL), a variation of ADSL where a modem can adjust the speed of the connection depending on the length and quality of the line. Using modified ADSL software, RADSL makes it possible for modems to automatically adjust transmission speeds, sometimes providing better data rates for customers located at greater distances from the central offices.
- High bit-rate Digital Subscriber Line (HDSL) is used for wideband digital transmission within a site and between the telephone company and a customer. HDSL is symmetrical (an equal amount of bandwidth is available in both directions) and HDSL can carry as much on a single wire of twisted-pair cable as can be carried on a T1 line (up to 1.544 Mbps) in North America or an E1 line (up to 2.048 Mbps) in Europe. The oldest of the DSL technologies, HDSL continues to be used by telephone companies deploying T1 Services at 1.5 Mbps and requires two twisted pairs.

- ISDN DSL (ISDL) is primarily geared toward existing users of ISDN. ISDL is slower than most other forms of DSL, operating at fixed rate of 144 Kbps in both directions. The advantage for ISDN customers is that they can use their existing equipment, but the actual speed gain is typically only 16 Kbps (ISDN runs at 128 Kbps). IDSL provides up to 144 Kbps transfer rates in each direction and can be provisioned on any ISDN capable phone line. Compared to ADSL and other DSL technologies, IDSL can be used at further distances from the central offices, and by users who are not served directly from the central office but through digital loop carriers and other technologies.
- Multirate Symmetric DSL (MSDSL) is Symmetric DSL that is capable of more than one transfer rate. The transfer rate is set by the service provider. Voice-over DSL (VoDSL) allows multiple phone lines to be combined into a single phone line with data-transmission capabilities.

SW56

Switched 56 (SW56) is a dial-up digital service provided by local and long distance telephone companies, which requires a DSU/ CSU for connection rather than a modem. A Digital (or Data) Service Unit/Channel Service Unit (DSU/CSU) is a pair of communications devices that connect an inside line to an external digital circuit. The DSU sends and receives signals while the CSU terminates the external line at the customer. A CSU may not be required in certain T1 ready communication devices. SW56 is the traditional data network in the United States using an analog signal with 56K bandwidth.

T1

T1 is a digital transmission link with a capacity of 1.54 Mbps over two twisted pairs of wires. One pair is used to transmit, the other to receive. T1 service accommodates 24 voice signals or any combination of voice and data signals up to 1.54 Mbps. A T1 line is plugged into the phone system for voice and into the network's router for data transmission. The 64 Kbps channels can be divided into any combination of voice and data transmission duties when a T1 link is configured. Initially designed for voice, T1 and T3 lines are now widely used to create point-to-point private data networks. Cost is generally based on the length of the circuit. E1 is the European version of T1.

T3

A T3 line is a dedicated phone/data connection supporting data rates of about 43 Mbps. T3 consists of 672 channels that each support a 64 Kbps data link. Each 64 Kbps link can traditionally support one voice conversation. T-3 lines are used mainly by Internet service providers (ISPs) and are sometimes referred to as DS3 lines.

ATM

Asynchronous Transfer Mode (ATM) is a form of data transmission that allows voice, video, and data to be sent along the same network. ATM is a key component of broadband ISDN having a high bandwidth, low delay, packet-like switching, and multiplexing technique. Information is divided among short, fixed-length packets called cells for transport, transmitted, and then re-assembled at their final destination. It is asynchronous because cells are not transferred periodically but are given time slots on demand.

Since ATM provides "bandwidth on demand", customers can be charged only for the data they send. It is best known for its ease of integration with other technologies and its sophisticated management features. ATM service works for applications that require bandwidth at speeds of 1.5 Mbps and higher. Because ATM divides data for transport into fixed-length, 53-byte cells, it supports high speed plus low delay, which means voice and video can run on the same infrastructure as data with no loss of quality.

Interoperability between the ATM equipment of different manufacturers and gateways to existing LAN/WAN standards mean maximum performance.

Cable Transmission

Cable Internet works by using TV channel space for data transmission. It can make your Internet access many times faster than with a conventional dialup system. Cable Internet works over the same hybrid fiber coax (HFC) networks that provide cable television service. In the case of HFC networks, coax cable is used within neighborhoods, and optical fiber often connects neighborhoods to central facilities. A cable modem is needed to access the Internet in this manner.

A cable modem has two connections: one to the cable wall outlet and the other to a PC or to a set-top box for a TV set. More complex than the telephone modem, the cable modem attaches to a standard 10 BASE-T Ethernet card in the computer. The wall outlet leads to a company cable line that connects with a Cable Modem Termination System (CMTS) at the local cable TV company office and can only send and receive signals to and from the CMTS. Cable modems cannot communicate with each other. In a business or commercial environment, the cable modem interfaces with a local area network (LAN) through an Ethernet hub, switch, or router and provides access to numerous users through a single cable modem.

Many cable operators are beginning to deploy high-capacity packet transport solutions over fiber rings connecting the CMTS units in their distribution hubs, such as Packet over SONET (POS), at up to 622 Mbps. In addition to the faster data rate, an advantage of cable transmission over telephone for Internet access is that it is a continuous connection.

Ethernet is the most common Local Area Network used by PCs. Ethernet is a specification for a LAN that is intended to carry 10 Mbps of data for use in the same building over high quality twisted pair wires or coaxial cable. There are 4 versions of Ethernet LANs:

10 Base-2   thinnet—coaxial cable and BNC connectors—180
    meters max—30 nodes
10 Base-5   thick Ethernet—thick coaxial cable—500 meters max–
    100 nodes
10 Base-T   twisted pair cable—RJ 45 connector—100 meters
    max—2 nodes
10 Base-F   fiber optic cable—2000 meters max—2 nodes

Fast Ethernet is IEEE standard 802.3u. Fast Ethernct raises the
communications bandwidth from 10 Mbps to 100 Mbps with only
minimal changes to existing Ethernet cable structures:

100 Base-TX   for use with Category 5 cable—RJ 45 connector
100 Base-FX   for use with fiber optic cable
100 Base-T4   for use with Category 3 cable—two extra wires

Other LANs include token Ring, Fiber Distributed Data Interface
(FDDI), and LocalTalk.

Frame Relay

Frame relay is an extremely efficient form of data transmission
frequently used to transmit digital information over local area
network (LAN) and wide area network (WAN) with variable connection speeds of 128 k up to 1.5 Mbps. Frame relay was originally
designed for use across ISDN interfaces using packets referred to
as frames. Today, it is used over a variety of other network interfaces as well. The technology was intended to be an intermediate
solution for the demand for high bandwidth networking. It is a
packet switching technology, which relies on low error rate digital
transmission links and high performance processors.

Frame relay is provided on fractional T-1 or full T-carrier
system carriers. Frame relay complements and provides a mid-
range service between ISDN, which offers bandwidth at 128 Kbps,
and ATM, which operates in somewhat similar fashion to frame
relay but at speeds from 155.520 Mbps or 622.080 Mbps. A Frame
relay requires a circuit to be installed by the phone company.

Fast Ethernet

Fast Ethernet is a local area network (LAN) transmission standard that provides a data rate of 100 megabits per second. It is commonly referred to as 100BASE-T. Workstations with existing 10 megabit per second or a 10BASE-T Ethernet card can be connected to a Fast Ethernet network using three types of physical wiring:

- 00BASE-T4 (four pairs of telephone twisted pair wire)
- 100BASE-TX (two pairs of data grade twisted-pair wire)
- 100BASE-FX (a two-strand optical fiber cable)

There is a vast scope of applications, and the various methods are mutating and evolving at a rapid pace.

Wireless Video Transmission

8

Wireless may be an old-fashioned term for a radio receiver, but there is nothing old-fashioned about wireless video transmission. Communicating digital images without the benefit of cables, cords, or wires is now almost commonplace. Today, the term is practically universal for the transmission of data between devices via radio frequency, microwave, or infrared signals.

ELECTROMAGNETIC RADIATION

An energy wave called a radio wave, which is generated by a transmitter, is a complex form of energy containing both electric and magnetic fields. This combination of fields cause radio waves to also be called electromagnetic radiation. All wireless data is carried by some form of electromagnetic radiation, and the most commonly used for wireless data communication include radio waves, microwaves, and infrared. Many remote video transmission systems used in security applications today are equipped

129

to operate wirelessly via radio frequency, microwave, or satellite link-ups in the case of a system interruption. These features are considered as backup or redundant systems that automatically take the place of the wired transmission systems should they terminate for some reason. Natural disasters such as wind storms, earthquakes, or floods provide an excellent example of the necessity of backup systems when land based communications are interrupted. In cases where traditional phone and cable communications systems are inoperable or unavailable, wireless communication capabilities are critical.

The electromagnetic field is used to transfer energy from point to point with an antenna as the source of these electromagnetic waves. Antennas are simply electronic components designed to send or receive radio waves. Energy is sent into space by a transmitting antenna, and the signal is then picked up from space by a receiving antenna. The design of the antenna is important because it determines the efficiency with which energy is transmitted. An efficient transmitting antenna must have exact dimensions. Characteristics are essentially the same for sending and receiving electromagnetic energy. This interchangeability is known as antenna reciprocity.

Antenna systems are required for all broadband wireless networks to operate. There are four basic styles or types of antennas used for broadband wireless systems: the sector (hub) antennas, flat-panel antennas, parabolic (dish) antennas, and dual-band antennas. A complete antenna system consists of a coupling device that connects the transmitter to the feeder and the feeder, a transmission line that carries energy to the antenna, and the antenna itself, which radiates energy into space. Factors determining the kind of the antenna used are the frequency of operation of the transmitter, the amount of power to be radiated, and the general direction of the receiving system.

The majority of antennas have evolved from two basic types, the Hertz and the Marconi. The basic Hertz antenna is 1/2 wavelength long at the operating frequency and is insulated from ground. The basic Marconi antenna is 1/4 wavelength long and is either grounded at one end or connected to a network of wires called a counterpoise.

Table 8-1 Frequency Levels

Extremely low frequency	(ELF, ITU band 2)	Below 300 Hz
Voice frequency	(VF, ITU band 3)	300–3000 Hz
Very low frequency	(VLF, ITU band 4)	3–30 kHz
Low frequency	(LF, ITU band 5)	30–300 kHz
Medium frequency	(MF, ITU band 6)	300–3000 kHz
High frequency	(HF, ITU band 7)	3–30 MHz
Very high frequency	(VHF, ITU band 8)	30–300 MHz
Ultra high frequency	(UHF, ITU band 9)	300–3000 MHz
Super high frequency	(SHF, ITU band 10)	3–30 GHz
Extremely high frequency	(EHF, ITU band 11)	30–300 GHz

Radio Waves

Radio waves carry information bearing signals that are either encoded directly on the wave by periodically disrupting its transmission (similar to telegraph), or the information is impressed on the carrier frequency by a process called modulation. Modulation is achieved by varying some characteristic of the wave. Amplitude modulation, known as AM, is achieved by manipulating the intensity or amplitude, and frequency modulation or FM is achieved by manipulating the frequency of the wave. See Table 8-1.

Infrared Transmission

Infrared data communications systems use infrared beams to carry data with light pulses. Another term for infrared transmission is beaming. With this type of system, video is superimposed onto an infrared beam by a transmitter and aligned to strike a receiver where the signal is output as a conventional composite video signal. The performance of infrared beams can be affected by weather and environmental conditions.

A wavelength ranging from 750 to 1,000,000 nanometers places infrared between microwaves and visible light in the spectrum. Infrared beams can quickly transmit data between two locations in the same room but it cannot carry information through or around walls or other obstacles. Most infrared communications systems are used to connect local workstations, PCs, and printers.

Microwave Transmission

Microwaves are commonly used in communications links spanning short distances, such as cellular telephone systems. These high frequency electromagnetic waves fall in a range between radio and infrared, with an approximate wavelength of one millimeter to one meter. Microwave links are capable of much farther transmission distances than infrared, and they are largely unaffected by weather conditions.

CELLULAR TRANSMISSION

Normally, a geographic region served by a cellular system via microwave is divided into areas called cells. Each cell has a central base station and two sets of assigned transmission frequencies—one set is used by the base station and the other by mobile telephones. This is where the name cellular comes from. To prevent radio interference, each cell uses frequencies different from those used by its surrounding cells. When a mobile telephone leaves one cell and enters another, the call is transferred from one base station and set of transmission frequencies to another by a computerized switching system.

A cellular telecommunications system consists of a portable or mobile radio transmitter and receiver, or cell phone, linked by radio frequencies to base stations. MTSO stands for the Mobile Telephone Switching Office that connects all of the individual cell towers, controls the cell sites, and manages all of the mobiles via a control channel. It is the link between the Public Switched Telephone Network (PSTN) and the cell sites providing the cellular customers connectivity to the standard phone network.

With cellular transmission, each cell phone identifies itself to the cellular system every time it places or receives a call. A cell phone's identity includes two unique values—the phone number that is assigned by the service provider, called a Mobile Identification Number (MIN), and an Electronic Serial Number (ESN). The ESN is a 32-bit binary number that cannot be changed.

Originally, American cellular telecommunications relied on an analog transmission system known as Advanced Mobile Phone Service (AMPS). These first generation cellular services essentially transmit voices as FM radio signals. Although AMPS is still in place, it is rapidly being replaced by digital technology. The most common of these are Time Division Multiple Access (TDMA), Code Division Multiple Access (CDMA), and Personal Communication System (PCS), a version of the European Global System for Mobile Communications (GSM). Frequency Division Multiple Access (FDMA) is used mainly for analog transmission. While it is certainly capable of carrying digital information, FDMA is not considered to be an efficient method for digital transmission. TDMA assigns each call a certain portion of time on a designated frequency. CDMA gives a unique code to each call and spreads it over the available frequencies, and FDMA places each call on a separate frequency.

The total number of mobile subscribers worldwide exceeds one billion and is growing. Many experts predict that number will double before the year 2010. Mobile wireless networks are continuously evolving and will be increasingly important enablers of new business and consumer applications including the transmission of digital video for the near future.

TDMA: Time Division Multiple Access

TDMA systems build upon the AMPS framework. TDMA may also be referred to as DAMPS, Digital AMPS, or US Digital. When the wireless industry first began to explore converting the existing analog network to digital, the Cellular Telecommunications Industry Association (CTIA) chose TDMA, in which three time slots are assigned to a single carrier frequency, allowing one channel to support multiple phone calls. In other words, several users can access a single radio-frequency (RF) channel without interference.

Both analog and digital data can be transmitted with TDMA systems because it is a dual mode wireless transmission method with bandwidth divided into common intervals for different data streams. The time slots used in each channel increase the amount of data that can be transferred over analog cellular systems. TDMA is also the access technique used in the European digital standard (GSM) and the Japanese digital standard, personal digital cellular. GSM systems use encryption to make phone calls more secure. GSM operates in the 900-MHz and 1800-MHz bands in Europe and Asia, and in the 1900-MHz band in the United States.

CDMA: Code Division Multiple Access

CDMA is a form of wireless multiplexing that is used to send data over multiple frequencies at once making it possible for many conversations to occur on a single channel. The technology is used in cellular telephone systems in the 800-MHz and 1.9-GHz bands. CDMA codes each digital packet (data) with a unique key that a CDMA receiver responds to. CDMA phones consume less power than analog phones because they are idle between bursts.

Variations of CDMA include B-CDMA, W-CDMA, and composite CDMA/TDMA. Developed originally by Qualcomm, CDMA is characterized by high capacity and small cell radius, employing spread-spectrum technology and a special coding scheme.

GSM-PCS Cellular System

The Personal Communication System, sometimes referred to as GSM-PCS Cellular System, is a TDMA digital system that changes voice and access information into digital data. GSM-PCS uses radio frequencies with approximately three times the efficiency of TDMA.

EDGE: Enhanced Data Rate for GSM Evolution

EDGE is a faster version of the GSM wireless service, designed to deliver wireless multimedia IP-based services and applications at theoretical maximum speeds of 384 Kbps with a bit-rate of 48 Kbps per timeslot and up to 69.2 Kbps per timeslot in good radio conditions. This enables the delivery of multimedia and other broadband applications to mobile phone and computer users. The EDGE standard is built on the existing GSM standard, using the same time-division multiple access frame structure and existing cell arrangements.

2G and 2.5G

Wide-area wireless technologies are commonly referred to by their generation of technology. As mentioned earlier, 1G refers to the original analog systems in America (AMPS). Second generation or 2G represents digital circuit switched networks. 2.5G is the evolution of the digital network to include packet-switched data.

3G Wireless

3G Wireless is an International Telecommunications Union (ITU) specification for cellular communications technology, also known as 3G, IMT-2000, International Mobile Telecommunications 2000, and Third Generation Wireless.

WIRELESS BROADBAND ACCESS TECHNOLOGIES

Wireless broadband access technology refers to high-speed wireless access services that provide bandwidths that exceed DSL and cable network technologies. Wireless Wide Area Networks, Wireless Local Area Networks, and Wireless Personal Area Networks all utilize complex communication techniques enabled by the increasing capabilities of digital signal processing.

There are several bits of jargon that would be useful to identify at this point to avoid confusion. For example, some services are considered to be fixed wireless. This means even though the data transmission is wireless, the stations are fixed rather than moving or mobile wireless. The term local refers to the signal's range limit. Point-to-point refers to a long-range wireless network between two points. A wireless network in which one point (the access point) serves multiple other points around it is known as point-to-multipoint. Indoor wireless networks are all point-to-multipoint.

Local Multipoint Distribution Service

Local Multipoint Distribution Service (LMDS) provides two-way Internet access using radio waves. It is a broadband point-to-multipoint communication system, developed as a wireless local loop, to be used in areas where installing physical cable is either impossible or cost prohibitive. LMDS uses a point-to-multipoint radio topology where a base station broadcasts to subscriber antennas. A transmission speed of several billion bits per second (gigabits) is possible along line of sight distances of several miles.

JPWL

Wireless communications offers its own particular set of problems that can make transmitting video data a challenge. To address this issue, the JPEG committee has established a new work item, JPEG 2000 Wireless (JPWL), as Part 11 of JPEG 2000 to standardize tools and methods to achieve the efficient transmission of video data over wireless networks. The JPWL system supports the functions of protection of the code stream against transmission errors, the description of the degree of sensitivity of different parts of the code stream to transmission errors, and the description of the locations of residual errors in the code-stream. JPWL has made JPEG 2000 very resistant to transmission errors, rendering it a good choice for wireless applications.

Wireless LAN (WLAN)

Wireless LANs provide high-speed data within a small region, such as a campus or small building. The standard, IEEE 802.11, specifies the technologies for wireless LANs. The standard includes an encryption method—the Wired Equivalent Privacy (WEP) algorithm. Most wireless LANs in the US frequency bands are located at 900 MHz, 2.4 GHz, and 5.8 GHz because an FCC license is not required to operate in these bands. The downside of this is that many other systems operate in these bands, causing interference between systems. There are two modes of WLAN.

Wireless Fidelity (Wi-Fi)

Wi-Fi wireless networking, also known as 802.11 networking, is a set of standards for wireless local area networks (WLAN) based on the IEEE 802.11 specifications. The IEEE 802.11 specifications are standards that specify an over-the-air interface between a wireless client and a base station or access point, as well as between wireless clients. The 802.11 standards can be compared to the IEEE 802.3 standard for Ethernet for wired LANs. Wi-Fi was originally intended to be used for wireless devices and LANs, but is now often used for Internet access. It enables a person with a wireless-enabled computer or personal digital assistant to connect to the Internet when in proximity of an access point called a hotspot.

Wireless mobile devices including portable computers, PDAs, and a variety of small wireless communication devices increasingly need to connect to corporate networks, perform database queries, exchange messages, transfer files, and even participate in collaborative computing. These wireless systems are achieving higher data rates to support Internet and other data-related applications, including the transmission of digital video.

Worldwide Interoperability for Microwave Access (WiMAX)

WiMAX is a more robust standard for high-speed broadband wireless delivery to laptops and desktops. The 802.16 standards are still being ironed out and products are not yet readily available as current development is concentrating on cost and power reduction.

WiMAX is the IEEE 802.16 technology that provides MAN (Metropolitan Area Network) broadband technology to connects IEEE 802.11 (Wi-Fi) hotspots. A hotspot is a connection point for a Wi-Fi network consisting of a small box hardwired into the Internet. The box contains an 802.11 radio that can simultaneously talk to up to 100 or so 802.11 cards. There are many Wi-Fi hotspots now available in public places like restaurants, hotels, libraries, and airports.

Metropolitan area networks are large computer networks usually spanning a campus or a city. They typically use optical fiber connections to link their sites.WiMAX provides high-throughput broadband connections over long distances at speeds of up to 75 Mb/sec. Signals can be transmitted up to 30 miles. Wi-Fi and WiMAX are complementary technologies.

802.11 is a family of specifications developed and maintained by the Institute of Electrical and Electronics Engineers. Most users know that the letters appearing after the reference denote speed, but they also define the distance the network can operate over, its security capabilities, and the frequency over which the networks operate. See Table 8-2.

Ultra-Wide Band (UWB)

UWB is a short-range radio technology that complements other longer range radio technologies like Wi-Fi and WiMAX and cellular wide area communications. It is a wireless technology that can transmit data up to 60 megabits per second and eventually up to one gigabit per second, used to relay data from a host device to other devices in the immediate area (up to 30 feet). UWB transmits ultra-low power radio signals with very short electrical pulses across all frequencies at once. UWB receivers translate short bursts of noise into data by listening for a familiar pulse sequence sent by the transmitter. Early UWB systems were developed mainly as military surveillance tools because they could "see through" trees and beneath ground surfaces. The greatest attribute of UWB devices is spectrum efficiency evidenced by its ability to operate

Table 8-2 802.11 is a Family of Specifications

802.11a	Applies to wireless ATM systems Enhanced data speed Frequency range 5.725 GHz to 5.850 GHz
802.11b	High data speed Low susceptibility to multipath-propagation interference Frequency range 2.400 GHz to 2.4835 GHz
802.11d	Allows for global roaming Particulars can be set at Media Access Control (MAC) layer
802.11e	Includes Quality of Service (QoS) features Facilitates prioritization of data, voice, and video transmissions
802.11g	Offers wireless transmission over relatively short distances Operates at up to 54 megabits per second (Mbps)
802.11h	Resolves interference issues Dynamic frequency selection (DFS) Transmit power control (TPC)
802.11i	Offers additional security for WLAN applications
802.11j	Japanese regulatory extensions to 802.11a specification Frequency range 4.9 GHz to 5.0 GHz
802.11k	Radio resource measurements for networks using 802.11 family specifications
802.11m	Maintenance of 802.11 family specifications Corrections and amendments to existing documentation
802.11x	Generic term for 802.11 family specifications under development General term for all 802.11 family specifications

Table 8-2 Continued

Wi-Fi	Originally created to ensure compatibility among 802.11b products Can run under any 802.11 standard Indicates interoperability certification by Wi-Fi Alliance
802.15	A communications specification for wireless personal area networks (WPANS)
802.16	A group of broadband wireless communications standards for metropolitan area networks (MANS)
802.16a	Enhancement to 802.16 for non-line-of-sight extensions in the 2–11 GHz spectrum Delivers up to 70 Mbps at distances up to 31 miles
802.16e	Enhancement to 802.16 that enables connections for mobile devices
802.1X	Designed to enhance the security of wireless local area networks (WLANs) that follow the IEEE 802.11 standard Provides an authentication framework for wireless LANs The algorithm that determines user authenticity is left open Multiple algorithms are possible
802.3	A standard specification for Ethernet Specifies the physical media and the working characteristics of the network
802.5	Standard specification for Token Ring networks

in an already overloaded spectrum without producing harmful interference.

WUSB

The universal serial bus (USB) technology is a popular connection type for computers and consumer electronic and mobile devices that can now provide the same functions of wired USB without the cables. The Wireless USB Promoter Group defines the specifications that provide standards for the technology.

Wireless Wide Area Network (WWAN)

A wide area network is a collection of local area networks connected by a variety of physical means. The Internet is the largest and most well known wide area network. Wide area wireless networks provide access to within a 20 mile radius with line of sight to an antenna, which is the limiting feature. Weather can interfere with traffic transmission.

Satellite Transmission

At the other end of the spectrum, so to speak, are satellite communication systems. Satellites in low earth orbit (LEO) circle the Earth at an altitude of 1242 miles. These low orbiting satellites obtain much clearer surveillance images and require less power to transmit data. The first communications satellite was NASA's Echo 1, an inflatable sphere that passively reflected radio signals back to Earth. The sphere successfully redirected transcontinental and intercontinental telephone, radio, and television signals. The success of this project proved that microwave transmission to and from satellites in space could be understood, exhibiting the potential promise of communications satellites.

Today's communications satellites relay information from one point to another via transponders. The transponder receives data at one frequency, amplifies it, and retransmits it back to Earth on another frequency. Communications satellites are usually geo-synchronous (GEO), which means they circle the Earth once every day at the exact same speed as the Earth rotates on its axis, making the satellite stationary to its position with the Earth.

Artificial satellites orbit the Earth from a vantage point that avoids the curvature-of-the-Earth limitations formerly placed on communications between ground-based facilities. By transmitting radio signals at high frequencies through the Earth's atmosphere, satellites can provide communications over great distances. Satellites are also beneficial as backup systems. In the event of natural disasters, a satellite may be able to receive and transmit information when traditional point-to-point methods are not functional.

Internet service via satellite is achieved by using a satellite dish for two-way communications. Upload speed is about one-tenth of its average 500 Kbps download speed. A key installation planning requirement is a clear view to the south, since satellites orbit over the equator area. Two-way satellite Internet uses Internet Protocol (IP) multicasting technology, which means up to 5,000 channels can simultaneously be served by a single satellite.

GLOBAL POSITIONING SYSTEMS (GPS)

Wireless data communication via satellites should not be confused with using satellites with GPS technology. GPS receivers use satellites to triangulate and calculate the device's longitude, latitude, and elevation, but they do not transmit signals. Inmarsat, the satellite company that provides roaming communications services to maritime vessels, military units, and aircraft, has launched a new generation of satellites that provide a wider range of global high-speed voice and data offerings. I-4 is the fourth-generation of Inmarsat satellites and will be the backbone of the company's Broadband Global Area Network, offering at least 10 times the communications capacity of the current network. The new satellites allow the company to offer data speeds of up to 432 K bits per second for uses such as video-on-demand, video conferencing, phone, e-mail, LAN, Internet, and intranet services.

Inmarsat satellites are used for peaceful communications, such as offering links between military units, but still require a high level of security. The company has its headquarters and satellite control room in downtown London, but has a mirrored site at an undisclosed location in the north of the city, which offers complete redundancy.

There are no limits to the uses for satellite transmission and no lack of imagination either. A new lightweight "communications suit" combines Inmarsat's latest Regional Broadband Global Area Network (BGAN) technology with state-of-the-art video monitoring tools. BGAN is an Inmarsat network comprising satellites and land earth stations that permits data speeds up to 432 Kbps. The suit, made by German and US companies, features a built-in still and video camera, personal computer, and a keyboard that is attached to the user's arm. Software developed by a Russian company send images, text, and video from the camera to the PC, and a wireless Bluetooth link provides the connection between the computer and the Regional BGAN satellite IP modem. Providing continuous coverage within the satellite footprint, it offers a secure 144 Kilobits per second shared channel, "always on" access to the Internet, and other IP-based networks across a large portion of the Americas, Europe, the Middle East, and India. Currently, billing is managed according to the amounts of data sent rather than time spent on line.

Terrestrial Microwave (TMW)

Microwaves are a form of electromagnetic radiation with frequencies ranging from several hundred MHz to several hundred GHz and wavelengths ranging from approximately 1 to 20 centimeters. Because of their high frequencies, microwaves are able to carry more information than ordinary radio waves and can be beamed directly from one point to another. On the other hand, these advantages make them more expensive than infrared links. There must be a clear "line of sight" between the transmitter and receiver of the microwave signal. For this reason, antennas are placed high so that the curve of the earth does not interfere with transmission. Microwave signals are largely unaffected by weather conditions.

WIRELESS SECURITY

Security vulnerabilities within 802.11 networks are well documented, and standards groups are moving quickly to solve these issues. 802.1x authentications, WEP, WPA, AES encryption, TKIP, MIC, and EAP-TLS are just some of the security fixes available.

802.1x

Standard 802.1x was approved by both the IEEE and ANSI in 2001. 802.1x is very scalable and supports a variety of authentication types. The IEEE 802.1x standard defines a port-based network access control to authenticate and authorize devices interconnected by various IEEE 802 LANs. IEEE 802.11i also incorporates 802.1x as its authentication solution for 802.11 wireless LANs. Based on industry standards from the IEEE, 802.1X authentication provide dynamic, per user, per session encryption keys instead of the static keys used in WEP or WPA. It enables centralized policy control, and a session timeout triggers re-authentication and new WEP key.

Wired Equivalent Privacy: WEP

Wired Equivalent Privacy (WEP) is a security protocol, specified in the IEEE Wireless Fidelity (Wi-Fi) standard 802.11b, designed to provide a level of security and privacy comparable to what is usually expected of a wired LAN. Wired LANs are generally protected by physical security mechanisms. WEP encrypts data transmitted over the WLAN.

Wi-Fi Protected Access: WPA

Wi-Fi Protected Access (WPA) is a standards-based, interoperable security enhancement that strongly increases the level of data protection and access control for wireless LAN systems. The technical components of WPA include Extensible Authentication Protocol (EAP), Temporal Key Integrity Protocol (TKIP), and 802.1X for authentication and dynamic key exchange.

Wi-Fi Protected Access 2 (WPA2)

WPA2 is the second generation of WPA security providing enterprise and consumer Wi-Fi users with a high level of assurance that only authorized users can access their wireless networks. It is based on the final IEEE 802.11i amendment to the 802.11 standard and essentially mandates the use of the AES encryption standard as an option alongside WPA's Temporal Key Integrity Protocol. WPA2 provides government grade security by implementing the National Institute of Standards and Technology FIPS 140-2 compliant AES encryption algorithm. WPA2 is backwards compatible with WPA.

AES: Advanced Encryption Standard

AES is a symmetric key encryption technique resulting from a worldwide call for submissions of encryption algorithms by the National Institute of Standards and Technology (NIST) in 1997 and completed in 2000. AES was selected by NIST as a Federal Information Processing Standard in 2001. The U.S. Government (NSA) announced in 2003 that AES was secure enough to protect classified information up to the top secret level.

EAP-TLS: Extensible Authentication Protocol—Transport Layer Security

EAP-TLS was created by Microsoft and accepted as RFC 2716: PPP EAP TLS Authentication Protocol. It is the de facto EAP used in 802.11i.

TKIP: Temporal Key Integrity Protocol

TKIP (Temporal Key Integrity Protocol) is part of a draft standard from the IEEE 802.11i working group. TKIP is an enhancement to WEP security, which adds a per-packet key mixing function to de-correlate the public initialization vectors from weak keys.

There are many advantages to using these types of wireless systems, including ease of setup and increased flexibility. The main disadvantage of wireless systems is their susceptibility to interference problems due to the fact that when using wireless transmission methods, a video signal is no longer on a dedicated run. This may lead to signal problems that can decrease video quality. However, the number of frequencies (channels) available reduce the probability of this happening, making wireless a popular mode for video transmission.

9

Examples of Digital Video for Security

The very word "security" can be ambiguous, meaning different things to different people. This is the definition given by Webster: *security n.*, 1. *Feeling secure, freedom from fear, doubt, etc.* 2. *Protection; safeguard.* Providing "protection" and "safeguard" have been a tall order for those in the security industry. Over the years many forms of security have evolved from those as simple as on-site guards to more complicated methods such as biometrics identification systems.

With the capabilities of digital video, security providers are now able to offer a more substantial version of the "protection" and "safeguard" referred to by Webster. The uses and benefits of remote digital video technology within the security industry have grown exponentially. Functions as diverse as border control, safeguarding employees, and monitoring hazardous materials make the applications of digital video immeasurable. Military, government, commercial, and private citizens all make up the spectrum of users of this technology. Following are examples of some common uses of the technology today.

CRITICAL INFRASTRUCTURE PROTECTION

Since September 11th, 2001 virtually everything is considered to be at risk, including energy supplies, water resources, bridges and tunnels, waterways, and airports. Improving the security at these and other potential targets has become a priority for our nation. Protecting our county's critical infrastructures and key assets involves a multitude of physical protection challenges due to the complex nature of the infrastructures and assets involved. Potential targets consist of a highly varied, mutually dependent mix of facilities, systems, and functions. Failure in one could conceivably begin a domino effect of consequences that could, in turn, negatively affect public health and safety, national security, the economy, and public confidence.

Scores of US dams are key components of other critical infrastructure systems, which provide water and electricity. There are approximately 80,000 dam facilities identified in the National Inventory of Dams. The federal government is responsible for roughly 10 percent of the dams whose failure could cause significant property damage or have public health and safety consequences. The remaining critical dams belong to state or local governments, utilities, and corporate or private owners. Current policies make dam owners principally responsible for the safety and security of their own facilities.

Idaho Power, an investor-owned, electric utility company headquartered in Boise, is taking positive action to prevent their facilities from becoming a target. That action includes the installation of digital video security equipment used to monitor three dams, power plants, and associated project facilities. The system is also used by Idaho Power's plant operators to view the areas below the dams prior to operating spill gates.

SCHOOL AND CAMPUS SECURITY

Schools and campuses have been implementing surveillance systems for years now, but digital technology has enhanced the abilities of these systems tremendously. For example, a school system in Washington State recently installed a digital video surveillance system that is linked with the local police and fire departments. When activated, the system gives police and fire department personnel visual access to most school areas from cameras placed at each entrance, facing the playing fields, in the library, and in each hallway. Live visuals of the hallways and outside are transmitted to responding individuals who may be outfitted with portable devices (both in vehicles and hand-held) able to receive the feeds. Response time is about the same, but knowledge of the situation, whether it is a fire or hostage situation, is immediately discernable. If a fire breaks out in one part of the school, firefighters are better prepared to know where and how to approach it.

The program is planned to tie in with the county's Amber Alert program, which deals with missing and exploited children. With this system in place, if a child is abducted from the school area, law enforcement officials can study outside camera feeds to track suspects or suspicious vehicles.

Vanderbilt University takes campus security very seriously and is committed to maintaining a safe, secure environment for students, faculty, staff, and visitors. Twenty-four hour foot and vehicle patrols, night transport/escort service, twenty-four-hour emergency telephones, lighted pathways/sidewalks, student patrols, digital surveillance systems, and controlled dormitory access (key or security card) work together to create a safe campus environment. The digital video system allows university officials to easily perform video searches and to e-mail digital video footage to the appropriate authorities when necessary. The university has saved time and reduced labor costs and incident response time as a result of upgrading to a digital closed-circuit television system.

School Coverage Areas

School security has developed extensively since so many incidents have gained large amounts of publicity in recent years. New schools are even being designed differently. For example, administrative offices are located so that the primary entrance can easily be seen at all times. Since many older schools cannot simply relocate the administrative offices, technology must be used to achieve the same result. By installing a camera at the primary entrance and a large wall-mounted monitor in the administrative office, the staff can easily see the entrance at all times.

An advantage of digital video systems for schools is the ability to access the system remotely in the event of a major incident. This means first responders can gain access to the camera system to view what is going on inside of the building, allowing them to direct their response to the appropriate areas and not waste efforts in areas where there is no immediate concern.

AIRPORT SECURITY

Airport security increasingly includes the electronic eyes of video surveillance technologies, which complement and support security personnel. Managing digital video surveillance over a computer network allows for the addition of thousands of cameras without the addition of staff to monitor the video data captured. A networked digital video management system is a perfect example of how to add camera "eyes" to an airport, or any facility, and automate the alarm triggers on the data captured so that security personnel are monitoring more of the facility without adding staff. Potential digital video tools will provide a return on investment not seen with previous surveillance systems while adding a greater sense of security at airports.

One of the most significant problems facing airports today are breaches at checkpoints and screening areas, even though these breaches are not necessarily caused by anyone with malicious intent. Unfortunately, when there are large numbers of people moving through a fast paced environment like an airport, someone may go through a checkpoint before they have presented proper ID or may gain entry through a screening area before they or their carry-on baggage have been properly screened. Such a security breach can create delays both for passengers as well as the air traffic network, which in turn can be responsible for tremendous economical loss to both the airport and airlines. A digital video surveillance system can be of tremendous value in reducing the problems associated with these types of breaches.

Another important surveillance challenge faced today is timely and accurate detection of potential threats from unattended baggage and other objects. There are many digital video systems on the market with features that detect non-moving objects, working opposite of the better known motion detection features.

A digital video system can offer more than additional security and surveillance to an airport; it can also provide some of the due diligence necessary in detaining suspects. With the right video system, the chain of events following a breach is expedited. Many different people may need to be involved in deciding a course of action. With a digital system, everyone concerned can instantly and positively identify an offender by viewing the same video recording of the event from predetermined locations around the airport. A physical description is immediately available and tracking of the offender can be viewed from said locations. This could allow airport officials to close off and search a very specific area, saving both time and money.

With digital video in place, airport security has the option of interfacing digital video data with other data sources, making the system more efficient, and more importantly, more effective. By converting video to a computer network, the door is open to activate software tools to recognize changes captured by the video data. With a networked digital video management system, advanced software tools have the capability to detect traffic patterns throughout the airport, ensuring that unscreened travelers are not entering secure areas. Key factors in an airport's decision to implement digital video should be high frame rate, open architecture for ease of integration, wide range of storage options, and user friendly system management.

NATIONAL MONUMENTS

Current global and political climates have increased the need for all public facilities to place a more concentrated emphasis on public safety and asset protection. As part of this trend, the National Park Service determined that an evaluation of Mt. Rushmore's security systems was necessary. The evaluation concluded that the original system installed was inadequate and should be expanded and modernized. Emphasis was placed on visitor safety and protection of the memorial as a whole.

A basic monitoring system with limited CCTV coverage was already in place and was incorporated into the new digital system. The new system provides a wide variety of security monitoring, using numerous types of technology, card access, and CCTV. All components are integrated together to make one seamless system. The foundation of the system is an NT server that communicates to remote control panels, workstations, and CCTV via fiber optic technology.

Every facet of building security—including access control, intrusion detection, video badging, and closed circuit television—is linked to a common database and controlled from a single operating platform. Security personnel monitor various input devices such as readers, keypads, door contacts, and request to exit switches, as well as control door locks, alarms, and other output devices. Special benefits of the new system include the ability to communicate to all buildings and sites via fiber optic cable for both monitoring and CCTV. Fiber optic lines were required due to the high incidence of lightning strikes in the area. The fiber optical lines, by nature, provide electrical isolation between system components around the site.

The CCTV system is integrated with the intrusion detection and duress alarms to provide instantaneous camera call-ups upon alarm, and the system allows National Park Service staff the ability to visually monitor areas more efficiently. Additional duress and intrusion detection alarms allow instantaneous notification of the Resource Management and Visitor Protection staff in case of trouble. Vehicular traffic is limited by the new system such that only authorized vehicles can reach sensitive areas of the memorial. Access to non-public facilities is controlled by proximity cards and PINs to exclude the general public from these areas. Perhaps the biggest concern at national monuments is the threat of someone leaving behind an explosive device. Proper video coverage of public areas within or around a monument can be used to watch for suspicious activity.

SURFACE TRANSPORTATION

The Federal Transit Administration National Transit Database reported a total of 132,293 criminal incidents related to surface transport in the year 2000, including 12 homicides. Data reported included approximately 450 of the largest transit agencies and only included incidents where arrests were made. The crimes reported included forcible rape, robbery, aggravated assault, larceny/theft, and motor vehicle theft. Public transportation such as buses, light rail, subway systems, and platforms not only require monitoring but also greater intelligence for preventative efforts. Escalation of an event can become acute in these environments because of the vast amounts of people that may be involved. Making public transportation safer and more secure for riders and reducing exposure of municipalities to liabilities is a goal shared by many. Transit agencies have scrambled to beef up passenger security by looking to new technology, including mobile video surveillance systems. These systems can monitor and record onboard events, collect footage from inside and outside, store operational data, and generally improve the quality and safety of public transportation.

The Santa Clara Valley Transportation Authority (VTA) of San Jose, CA has cameras on up to 630 light rail vehicles and buses throughout the Santa Clara County area. The Charlotte Area Transportation System (CATS) Authority project includes systems on 285 buses. Each bus is fitted with a six-camera system that captures both video and audio data. CATS uses the system for three main functions: accident investigation, driver training, and enhancing security on the fleet. Two cameras tape the exterior of the vehicle—one views the area in front of the bus and the other down the passenger boarding side. Inside the bus, cameras are positioned to see boarding, fare payment, and areas down the bus and toward the rear of the bus. San Francisco Municipal's system includes coverage that helps deal with issues such as vagrants using buses as shelter, vandalism, unauthorized access to buses while in the depot area, and flagging of specific incidents and/or passenger counts for insurance review.

During a testing phase on VTA vehicles in California, over 200 arrests were made for vandalism alone. The system continuously records and holds approximately 80 hours of video. When an incident occurs, drivers are able to indicate that a recording should not be overwritten and security personnel can pull video nightly or go on site to an incident and view video for fast analysis.

THE HEALTH CARE INDUSTRY

Many security professionals have horror stories such as stolen cars, employee attacks, vandalism, and expensive equipment being stolen. Babies missing from the maternity ward is one that most security managers do not have to worry about. University Health Care expanded their existing security system and integrated it into a PC based system that allows them to monitor the entire hospital. Not a moment goes by that vulnerable spots in the facility aren't being monitored and recorded. In addition, the new camera system does more than just watch, it reacts. When the camera senses motion in an area where there shouldn't be any, it alerts the appropriate personnel.

The original job, completed in August 1999, consisted of a closed circuit surveillance and panic alarm system. Since then, an extra parking deck with nine additional pan/tilt/zoom cameras and several panic alarms has been added to the system. Since the installation of the parking deck cameras, University was able to record a car theft and reduce loitering instances. Also, employees of University Health Services have a higher sense of safety while at work. University now has the option to replace their aging access control system and integrate it with the new closed circuit surveillance and panic alarm system. This will allow University to have one completely integrated security management system.

FAR AWAY PLACES

A new digital system has been set up at the Caribe Hilton Hotel and Casino of Puerto Rico primarily to monitor access/egress to and from hotel facilities. All major entrances and exits are monitored, and secondary stages of surveillance include access points to ballrooms, meeting rooms, the main driveway, and all public areas of the lobby, including guest elevators at the lobby level. "Back of the house" sensitive area surveillance includes the loading dock, service corridors, accounting office, general cashier, general storeroom, and drop safe room. Finally, the system covers the inside of all service elevators.

Full 24 hour surveillance of the above areas serves as a deterrent for criminal activity throughout the property, simultaneously providing guests and associates with the feeling of being in a safe and secure environment. It also provides the security department with a powerful crime prevention tool and the ability to record accidents and/or incidents that may occur throughout the property. Damage to cargo and service elevators, in most cases done by delivery persons and outside contractors, has been eliminated. Since the cameras' installation, the hotel has only experienced two incidents of this nature. In each case, the events were caught on video and the hotel was able to identify those responsible and collect payments for the damages.

SECURITY AT THE SECURITY CONVENTION

The folks who attended the 2002 ASIS seminars and exposition at the Pennsylvania Convention Center were lucky enough to enjoy the award winning blend of classic style and modern technology brought together within the 1.3 million square foot facility. What they may not have known was that a new digital video system was installed at the Pennsylvania Convention Center as part of a security upgrade. The primary reason for the system is to ensure the safety of the staff, convention center attendees, and clients as well as to combat typical convention center thefts: laptops, bags, and purses. Spokesman for the Center said that almost all incidents prior to installation of the new system went unresolved by local authorities and/or convention center security personnel. The cataloging of tapes was highly ineffective and investigation was difficult. Using the old recording system, there was no indication that the analog recorders were ever used or useful for post-incident investigations. Once the new digital system was installed, all incidents that occurred within camera view have been resolved, including five major incidents. Major incidents would involve items valued in excess of $50,000 per incident. In fact, the Convention Center has received several letters from the Philadelphia Police Department commenting on the usefulness in providing evidence for prosecution of suspects involved in criminal activity in and around the facility. The system has also proven to be a highly successful liability mitigation tool.

RETAIL SECURITY

As part of Home Depot's growth philosophy, the Atlanta based retailer recently launched a $250 million store remodeling program with the intention of streamlining store operations and redeploying gained efficiencies to improve customer service. Included in the overhaul is the rollout of a new digital video surveillance system that will provide a safe environment for customers and associates, aid in combating identity theft and credit card fraud, reduce in-house shrinkage, cut down on cash register errors and other fraudulent actions, and work to eliminate shoplifting.

The new digital surveillance technology integrated with Home Depot's POS system enables the retailer to view individual transactions as well as provide real-time, remote monitoring of store operations from the company's district offices or its headquarters in Atlanta. The platform is a software-based tool that stores, organizes, and delivers digital images collected by surveillance cameras.

Approximately 100 cameras per store will monitor checkout stands, receiving docks, store aisles, and various other parts of its operations, including parking lots. With the new system, Home Depot can actually e-mail law enforcement agencies with digital photos or action sequences or evidence can be downloaded onto a CD for transport or storage. While Home Depot is certainly not the first retailer to use digital video surveillance, it is one of the first of its kind to use a coordinated digital system throughout all store operations. The end result will be approximately 40,000

Figure 9-1   Dome Camera In Retail Facility

video cameras in use to better monitor stores, reduce shoplifting, and deter fraud.

It is an unfortunate fact that the "convenience" of convenience stores often accommodates both the consumer and the criminal. For the consumer, convenience stores provide easy access to groceries, prepared foods, gasoline, and other services. From the perspective of the criminal, these stores offer easy access to money. Convenience stores, like other establishments that deal with merchandise and money, also offer the occasional temptation for theft to employees. Digital video systems are making this kind of crime much harder to conceal.

The use of digital video is not limited to security purposes. With continuous digital recording, important marketing information can be gathered and disbursed to appropriate management personnel or even consultants. In the case of a retail application, pertinent data can be gleaned such as peak business hours, customer profiles, the attractiveness and attention getting quality of displays or product arrangement. Inventory levels can be assessed and addressed from off-site and general facility appearances appraised. Improved processes, marketing strategies, customer satisfaction and retention, and improved sales are just a few benefits possible.

With networked digital video, site management takes on a new dimension. Not only can a store manager remotely inspect the appearance of stores, displays, and inventory but the behavior and presence of staff can also be monitored. If a store manager suspects a cash register has been tampered with after closing, he or she can retrieve and review video images of that cashier station between certain hours and only when the register was opened. In another example, a truck entering an off-limits loading area can deliver an alert message containing video images.

Evaluations can more easily be made about the effectiveness of the site layout as well as the efficiency of the staff to customer ratios. Situations such as unsecured doors, wasted utilities, low stock levels, erroneous product shipments, and inefficient work-flows can all be documented and corrected. Improvements in personnel conduct and store appearance are immediate, and savings related to improved site management alone should result in a significant return on investment. Surveillance cameras have been around for a long time in the retail industry, but digital technology has increased its business benefits. Many liability suits can be won with the evidence provided by a continuously recording digital system if you have the ability to hand over authentic video information.

CARS ON VIDEO

A provider of some of the world's finest automobiles, located in the heart of Coral Gables, Florida is home to a collection of cars, ranging in price from $30K to more than $200K. A state-of-the-art, three-story, 350,000 square-foot facility called The Collection houses seven world-class auto franchises under one roof including Maserati, Aston Martin, Ferrari, Porsche, Jaguar, Audi and Lotus. The building is secured by an extensive access control system, including biometric readers and integrated digital CCTV and communications systems. The CCTV portion contains 105 color cameras plus five integrated PTZ color dome cameras, with five monitoring and control locations. The entire system is networked with five workstations that control all access points to the facility, lock and unlock showroom doors on schedule, as well as control the building elevators.

CITYWIDE SECURITY

City of London, Ontario turned to digital video technology as a means of crime prevention and as a means to verify and respond to criminal activities or other emergencies.

The plan officially began when a committee of elected public officials, members of the city police department, and chairman of a citizens committee initiated the "London Downtown CCTV Project" to deter future crime in the downtown location. After two and a half years of preparation, the cameras went live on November 9, 2001.

The project is currently comprised of cameras equipped with pan, tilt, and zoom capabilities installed at major intersections in the downtown core of the city. The cameras are monitored 24 hours a day, 7 days a week at city hall, and can also be viewed and controlled from police headquarters. The main function of the cameras is to improve community safety, crime prevention, and the desirability of the city as a place for shopping, business, and leisure.

This system is unique in the fact that all controls and video images are transmitted to the control station via wireless links. This was necessary due to the lack of fiber optic or copper wire available at the camera locations. Cameras are suspended from traffic light poles on steel arms that also hold omni directional antenna transmitters. These transmitters send the video and control signal to a receiver atop a centrally located high-rise office tower from which the signals are re-transmitted to city hall. At the camera location, the video signal is converted to a digital signal so that it can be transmitted over a wireless network. All cameras are being digitally recorded to provide visual evidence if required.

Preprogrammed tours are set up to view desired areas of the streets. When an incident occurs or suspicious activity is noticed, the operator will assume manual control and track the occurrence. The cameras are able to scan 360 degrees in less than a second, reading license plates at hundreds of feet and zooming in on a designer logo on a shirt or a tattoo on an arm. If police intervention is required, headquarters is alerted to view the camera and/or respond with officers. In one case, three men suspected of beating another man were identified and charged because they walked by security cameras outside police headquarters minutes after an assault a few blocks to the east.

A gas station robber was caught on video. The video was released to *The Free Press* and New PL TV and within 24 hours, a suspect was identified and charged with six robberies. Over all, this system leaves the City of London with a greater feeling of security and the police with an instant picture of incidents such as fights, demonstrations, vandalism, and robberies.

GAMING SECURITY

Gaming has become a multi-billion dollar industry and it is inevitable that wherever there is an exchange of large amounts of money, there is a need for security measures. Digital video systems provide more than just security and surveillance to casino, hotel, and resort management; they also provide some of the intelligence necessary to making business decisions about how the facilities are run. In a casino environment, surveillance is typically a reactive endeavor. Changing from a reactive to a proactive system allows casino security to be alarmed on an event, and gives them access to video data in seconds.

These tools can help to deter criminal activity such as card counting while simultaneously monitoring all areas of the casino. A small but growing number of casinos across the nation are utilizing even a more sophisticated surveillance tool: facial-recognition software and databases. With this software in place, a casino can look for known offenders automatically, record a problem guest's face for future reference, and even track high-roller guests throughout the facility, providing them with exceptional customer service. Gaming establishments all across the nation are realizing that the latest digital surveillance technology can better protect them against customer fraud, employee misconduct, and fraudulent claims.

FINANCIAL INSTITUTIONS

Financial institutions usually require an open architecture digital video solution that can be installed on a network and also provide a method to quickly send video data to law enforcement agencies. Another requirement is a system that marries into existing teller environments and the ATMs. Digital video technology provides a perfect solution.

Networked digital video can provide on-site capture and off-site retrieval of video and security and remote accessibility allows retrieval and administration over LAN/WAN, local, or modem connectivity. In addition, networked digital video systems can provide instant access for live monitoring of ATM atriums or in cash counting areas, as well as quick access to stored video data. It can also be configured to send alarm messages triggered by specific events such as video motion detection, digital inputs, and camera faults. When an event occurs, a message is sent with the time of occurrence and a brief description.

SECURITY IN THE BAYOUS

At the mouth of the Mississippi River on the southernmost tip of Louisiana lies some of the best fresh water, salt water, and offshore fishing in the world. Cypress Cove Lodge and Marina is located in this fisherman's paradise. Not only will you find fishing, hunting, occasional alligators, and scenic beauty at Cypress Cove, but also state of the art security. The Cypress Cove complex consists of not only the lodge but also a full service marina and marina store. The facility also provides 196 dry boat storage spots and 140 wet slips. The boat storage buildings, the store, and lodge all require security surveillance.

A remote digital video surveillance system was installed as an adjunct to the 24 hour guards in place. A system with seven cameras and one DVR is currently in place at Cypress Cove, which the owners use to keep a close eye on employees and the general management of the facility. There is a camera view of the fuel pump area so that they can be alerted to chemical spills and assign responsibility if they occur.

SECURING BUILDING SITES WITH SPECIALTY SURVEILLANCE EQUIPMENT

Construction sites are often isolated and devoid of inhabitants on holidays, weekends, and evenings, making them good targets for thieves who want to pick up building supplies unnoticed and unpaid for. In recent years, contractors nationwide have reported a wave of thefts of everything from standard building materials, tools, and appliances to heavy equipment and vehicles. These construction site thefts are resulting in millions of dollars in losses. The upsurge in thefts is blamed on a building boom and the rising price of materials. In Southern California, a unique digital video system with night vision and license plate recognition technology is used by one builder to assure that acts of material or equipment theft are likely to be captured on video. By producing clear digital images of the culprit in either day or night time, the theft is documented. By retrieving the license plate number of the vehicle involved, the thief is captured. After an incident of theft or vandalism, personnel who are trained in finding evidence of the incident as quickly as possible retrieve the stored digital data. Copies of the pertinent digital information are given to the customer along with a formal report of the incident and still shots from the video images obtained. In most cases, the information is given to local authorities who, with a clear license plate number as evidence, usually have no problem locating and convicting the perpetrators.

As is so often the case, the advantages of digital technology are its multitude of uses. Several customers have some form of broadband transmission, which makes images available for view from various locations on site or off.

10

Pieces and Parts

We know that digital compression is a method of converting information to a format that requires fewer bits and can be reversed to a close approximation of its original state once transferred to a new location. In the case of CCTV, this means that video can be digitized to a smaller form and therefore transmitted at a quicker speed. An important fact to remember when considering cameras for CCTV applications is that the final resolution quality will only be as good as the weakest link in the system. In other words, if you have high resolution capability from the camera but your monitor provides a lower resolution, the monitor will determine the resolution you receive. Do not make the mistake of assuming that because you have high resolution capable cameras, you will automatically receive high resolution images. Just like an automobile, all parts of the CCTV system must provide equal or similar performance levels to achieve the overall goal. A superior motor on a car with bad tires will not get you very far.

CAMERAS

Traditional analog cameras convert light intensities into images on light-sensitive film or tape. Digital cameras convert light intensities into discrete numbers for storage on an electronic medium, such as a hard disk or flash disk. Like their analog counterparts, a digital video camera captures a continuous stream of information, and a digital still camera captures a snapshot.

The job of the camera is to provide information from the site via electronic signal, cable, or phone line to the viewer. The two basic camera types are tube and chip. The use of the tube camera has gradually decreased despite its lower cost and higher resolution.

Advantages of the chip camera include fewer maintenance requirements, better durability, longer life, resistance to lag and bloom, and smaller size. The chip camera also produces a digital signal that allows for freeze of action and digital motion detection.

The camera location and environment will play an important role in deciding on the type of camera you use. Placement of the camera is also a determining factor, especially when it comes to the evidentiary value of the images captured. Variations such as whether the camera is located inside or outside, temperature extremes, light levels and how the light levels may change are all factors whether you are using analog or digital systems.

Charged Couple Device (CCD)

Tube cameras used to be the most commonly used type because they are relatively inexpensive and well suited to most indoor uses. They can be susceptible to damage from very bright lights, making them less desirable for outdoor use. The Charged Couple Device (CCD) cameras, sometimes called chip cameras, have moved up in popularity as prices have come down.

A chip camera works as light passes through the lens and forms an image on the CCD. The CCD is a semiconductor grid of thousands of pixels that produce an electric charge according to the color and brightness of light passing through it. It is an integrated circuit containing an array of linked, or coupled, capacitors. Under the control of an external circuit, each capacitor can transfer its electric charge to an adjoining capacitor.

Light hitting the surface of the CCD causes it to free electrons to move around and accumulate in the capacitors, providing a black-and-white image from the light that falls on each pixel. Repeating this process converts the entire contents of the array to a varying voltage that is sampled and stored. Stored images can then be transferred to a printer, storage device, or video display. Advantages of the chip camera include fewer maintenance requirements, longer life, resistance to lag and bloom and resilience to shock or vibration. They are also smaller and require less power to operate. The chip camera also produces a digital signal that allows for freeze of action and digital motion detection. CCDs are characteristically sensitive to infrared light.

Infrared Cameras

Though technology has not yet achieved X-ray vision, infrared and advanced thermal imaging runs a close second. The ability to see in the dark, through smoke, fog, and certain natural obstacles such as foliage brings infinite advantages for security, military, and fire fighters. Infrared cameras are not dependent upon digital technology, but the advances in digital transmission and storage increase their value tenfold.

Adding infrared capabilities to a video surveillance infrastructure provides both early detection and full visibility irrespective of the prevailing light levels or weather conditions. Infrared radiation is heat radiation generally produced by anything with a temperature above 10 degrees Kelvin (a unit of absolute temperature equal to 1/273.16 of the absolute temperature of the triple point of water, which is equal to one Celsius degree). It has many of the same properties as visible light, such as being reflected or refracted. Thermal images are produced primarily by self-emission and by emissivity differences. Emissivity is a measure of how much radiation is emitted from the object. Normally, object materials and surface treatments exhibit emissivities ranging from approximately 0.1 to 0.95.

A highly polished (mirror) surface falls below 0.1 while an oxidized or painted surface has much higher emissivity. Oil-based paint, regardless of color in the visible spectrum, has an emissivity over 0.9 in the infrared. Human skin exhibits an emissivity close to 1. Non-oxidized metals represent an extreme case of almost perfect opacity and high specular reflectivity, which does not vary greatly with wavelength. Reflection off of smooth surfaces such as mirrors or a smooth surface body of water is known as specular reflection, while reflection off of rough surfaces is known as diffuse reflection.

Consequently, the emissivity of metals is low—only increasing with temperature. Vehicles that have a low-level emissivity, such as automobiles, tanks, or aircraft, will normally emit thermal radiation from their energy source while operating, making them visible to thermal imaging devices. In other words, the hot areas such as the engine and the exhaust system will be visible. The same theory is true for vehicles that have recently shut down because their engines are still warm.

Thermal Imagers

Thermal imaging is considered long range infrared and does not require any additional lighting. Typically, the image produced is a black-and-white image, where the hotter objects are whiter and the cooler objects are darker. The heat sensing abilities of thermal cameras allow them to easily identify intruders and other security breaches at night. Thermal imaging has been proven to be a successful solution for common security needs such as:

- vision at night where lighting is undesired or unavailable
- surveillance over waterways, lakes, and ports where lighting options are impractical
- surveillance in challenging weather conditions

Thermal imaging devices extend normal vision by making thermal heat visible. For example, burning charcoal may not give off light, but it does emit infrared radiation, which is felt as heat. With thermal imaging technology, the ability to see the burning coals would be available even if the receiver (either human or mechanical) is too far away to actually feel the heat or actually see the coals.

Image Intensifiers

There is sometimes confusion between thermal imaging technology and image intensifiers or I^2 technology because the term "night vision" is often applied to both. Night vision can be achieved by intensifying the small amount of light present, even at night, from the stars and the moon. A device based on this principle is called an image intensifier (I^2) or starlight scope (SLS).

Image intensifier uses have evolved from nighttime viewing to fields including industrial product inspection and scientific research, especially when used with CCD cameras (intensified CCD or ICCD). I^2 devices offer daylight-like conditions to the user but do not mean the user will be able to detect objects normally seen in daylight conditions.

An image intensifier does not work in total darkness. It does, however, create a more realistic image than night vision because it reveals the same type of image that the human eye sees. Not based on temperature, I^2 technology relies on amplifying available light to detect objects and is often ineffective in extreme low-light situations such as heavy overcast. Another disadvantage of I^2 technology is blooming, which can occur if high intensity light in the field of view saturates the sensor's picture elements.

Digital Signal Processing/Digital Signal Processor

The introduction of DSP in security surveillance cameras has increased the flexibility of uses and the quality of the color image.

DSP technology offers more consistent picture quality over a wider range of lighting conditions. They can also provide features such as programmable intelligent backlight compensation, Video Motion Detection, remote set-up, and control and on-screen menus, making them a good choice for complex surveillance conditions. Digital video should be displayed at least at the same resolution as the camera sending the images to avoid distortion.

DSP can either refer to digital signal processing, the term used for processing signals digitally, or digital signal processor, which is a type of microprocessor chip. A digital signal processor is a programmable device and digital signal processing is exactly what its name suggests, a method of processing signals digitally using mathematical techniques to perform transformations or extract information. A DSP chip is designed to carry out processing of up to tens of millions of samples per second.

Digital signal processing takes real-time, high-speed information such as radio, sound, or video signals and manipulates it for a variety of purposes. The goal of digital signal processing is to use the power of digital computation to manage and modify the signal data. DSP systems translate analog signals into a digital approximation, but DSP used in cameras do not convert digital signals to analog signals.

LENSES

Several questions influence the choice of a lens, which determines the area viewed by the camera and makes adjustments based on light conditions. Will the camera be viewing a wide area fairly close to the camera? Will images from a narrow area far away be important? What are the lighting conditions, and will they change?

Format is the size of the imaging area on the tube or chip and the term that describes the size of the lens to be used for various image requirements. The focal length of the lens along with format determines the field of view captured. Focal length of the lens determines the field of view. A short focal length will produce a wide view and a long focal length, a short view. Lens speed is the light collection ability of the lens. The iris diameter is measured in f-stops: $f/1$, $f/1.4$, $f/2$, or $f/2.8$. Lower numbers are faster and let in more light, while the higher numbers reduce the light. It is necessary to understand the expectations of the camera and its environment before choosing a lens.

C-mount and CS-mount

There are two kinds of lenses commonly used with high-speed video equipment, and they refer to the two standard mountings used for connecting lenses to cameras. One is called C-mount and the other one is called CS-mount. C-mount lenses are used for larger image sensors. The CS-mounts are used for the more compact image sensors.

The difference between the two standards has to do with focusing. Every lens is designed to be a certain distance from the imaging element, whether that imaging element is a video sensor or film. This distance specifies the focal plane. C-mount lenses are designed for a focal plane distance that is 5mm greater than the same dimension for CS-mount lenses. In other words, C-mount lenses are designed to be 5mm farther from the sensor than CS-mount lenses.

If you have a C-mount camera, you must use C-mount lenses. If you have a CS-mount camera, CS-mount lenses work fine with your camera. C-mount lenses will work with your CS-mount system, but only with an adapter. The adapter is actually a spacer with both male and female 1–32 threads that moves a lens 5mm farther from the camera. A lens creates an image of a certain size on its focal plane. That size is part of the lens specification and is expressed in terms of the sensor diagonal dimension. It is possible to use a lens designed for a larger sensor (to use a 1-inch lens with a 2/3- or 1/2-inch sensor) but using a lens designed for a smaller sensor (to use a 1/2-inch lens with a camera with a 1-inch sensor) will bring unsatisfactory results, specifically a reduction of resolution and image size.

New lenses are available with a choice of manual DC iris and have a variable focal length setting. There are versions with 4-pin and 6-pin connectors for DC-controlled iris. The advantage to system designers or installation engineers is that they no longer need to calculate the field of view and the required focal length in advance. A rough estimate is sufficient. Upon installation, the focal length of the selected lens is manually adjusted to suit the situation, before focusing and setting the iris to meet ambient lighting conditions.

Varifocal

Varifocal lenses are the most flexible for applications requiring a wide range of focal lengths. Focal length adjustments are made by turning a dial. A limited number of varifocal lenses will cover a wide range of applications, which would have required a much larger number of lenses with fixed focal length.

Auto-Iris Lenses are designed for outdoor use or any applications with variable lighting conditions. They are available in C or CS Mounts from super-wide angle to telephoto (depending on the application use), DC and Video types. The DC type is more economical and designed for the newer CCD cameras, which incorporate ALC (Automatic Level Control) circuitry of the camera.

As in every technology, many pieces determine the success of the whole, and this is true in video technology as well. Lens determination is an essential component of a system regardless of its objective. It will be well worth the time involved in choosing an appropriate lens for your application in order to achieve the success desired.

ILLUMINATORS

When cameras are used outdoors or in low light areas such as parking garages, there is not always enough light to provide a recognizable image. Adding lights may be an option but is not always possible. When this is the case, other devices may be used.

Infrared illuminators transmit infrared light, which can be used to enhance the video quality. Infrared light, while not visible to the human eye, is within the spectrum of useable light for cameras. With monochrome cameras, the light emitted from an IR illuminator provides enough light to enhance the picture quality.

Most color cameras utilize an IR filter, rendering the illuminator ineffective. If an IR filter were not used with color cameras, the image colors could be drastically affected - particularly during the day. For example, on a hot summer day a black car would be much hotter than a lighter colored car. The heat given off by the surface of the black paint causes the video image to appear as a much lighter color.

SWITCHING EQUIPMENT

In many cases involving CCTV surveillance, a user will want to view more than one camera at the same time. Adding a monitor for each camera (which is what casinos used to have to do) can be extremely expensive in and of themselves as well as in regards to the space needed to mount them. Switchers allow a user to manipulate several camera views on one monitor. There are several options for switching available.

Bridging Switchers

A bridging switcher can display a single camera on one monitor and at the same time display sequencing cameras on a second monitor, making it possible to view a continuous sequence of cameras on one monitor and one specific camera on a second monitor.

Multiplexers

Multiplexers have been around for quite some time and are used to consolidate several communication channels into one channel of data. They allow an operator to view images from up to 18 cameras on one screen. Today there are dozens of brands and configurations to choose from. The multiplexer allows multiple signals to be transmitted simultaneously across a single physical channel and process the output from multiple cameras to a monitor or recording device, providing the ability to display the images from multiple cameras on one monitor. Images can be reviewed in a variety of screen formats, but all have to be played back through the multiplexer.

Most multiplexers are either simplex or duplex. Simplex records exactly what is seen on the monitor. A duplex records all camera images, regardless of what is being viewed on the monitor, and allows for review of any particular camera image by itself. Duplex multiplexers also allow the viewer to review recorded images without hampering the monitoring or recording of current scenes. A duplex multiplexer is useful for applications where monitoring is a key part of the system function and video tapes need to be viewed regularly.

Other multiplexers offer priority recording using motion detectors and video motion detectors. There is a distinction between the two: motion detectors respond to any change within its radius of detection such as temperature, pressure, or sound. Video motion detectors respond to any change in pixels on the screen.

Interleaving is another advanced feature that prioritizes sequential viewing based on a high activity area or detected motion. The priority view is seen more frequently in a sequence of views than other camera views of lesser importance.

Matrix Switchers

A matrix is a logical network configured in a rectangular array of intersections of input and output channels. A video matrix switcher is a device capable of interconnecting many components in a variety of combinations. This switcher allows any one of the inputs to be switched to any one or all of the outputs. In other words, it can switch more than one camera, VCR, video printer, or other similar device, or to more than one of these devices at once.

MONITORS—MORE THAN MEETS THE EYE

We are no longer limited to the constraints of the human eye. Now we can literally take a live or moving scene and transport it to another room, another state, and even another planet using the magic of digital video. Enter the monitor.

The resolution of a monitor is indicative of how many pixels it will display. Generally, more pixels mean better resolution. The number of colors that can be displayed at one time is limited by the amount of video memory installed in a system. Another classification of monitors depends upon the type of signal they accept, i.e., analog or digital. This can be a confusing because calling a monitor analog or digital refers to the type of color signals it uses although it may actually have either analog or digital controls. The size of the display area will affect the resolution in that the same pixel resolution will look better on a smaller monitor and fuzzier on a larger monitor because the same number of pixels is being spread out over a larger number of inches. A good example of this phenomenon can be seen by using a copy machine to repeatedly enlarge a picture. Your first few enlargements will look pretty good, but eventually the larger the picture the less definition it will have.

The relationship of width and height of a video image is called its aspect ratio. When an image is displayed on different screens, the aspect ratio must be kept the same to avoid stretching in either the vertical or horizontal direction.

CRT Monitor

Cathode-ray tube (CRT) technology has been used in most televisions and computer display screens until recent advances in flat screen technologies have made these more easily available and affordable. A CRT works by painting an electron beam back and forth across the back of the screen. Each time the beam makes a pass across the screen, it lights up phosphor dots on the inside of the glass tube, which then illuminates the active segment of the

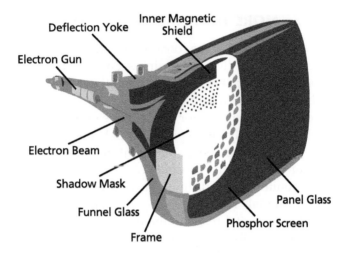

Figure 10-1   CRT Monitor

screen. Phosphors are chemicals that produce light when excited by electrons.

Monitors can actually be affected by where they are located on the Earth, because the Northern and Southern hemispheres have different magnetic fields. CRT monitors are manufactured specifically for whatever hemisphere they are going to be used in. These magnetic fields can have an effect on CRT monitors because they work by moving electron beams back and forth behind the screen. LCD monitors are not affected by this phenomenon. CRT monitors cost less and produce a display capable of more colors than LCD monitors do. See Figure 10-1.

LCD Monitor

A liquid crystal display is made up of an electrically-controlled light-polarizing liquid trapped in cells between two transparent polarizing sheets. The polarizing axes of the two sheets are aligned perpendicular to each other, and each cell is supplied with electrical contacts that allow an electric field to be applied to the liquid inside. LCDs are non-organic, non-emissive light devices—they

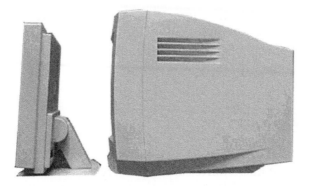

Figure 10-2   LCD Monitor and CRT Monitor Compared

do not produce any form of light but instead block light that is reflected from an external source. Some users report lower eye-strain and fatigue due to the fact that LCD displays have no flicker. One of the biggest advantages of LCD monitors is that they are compact and lightweight. See Figure 10-2.

The Digital Display Working Group (DDWG) is an open industry group lead by Intel, Compaq, Fujitsu, Hewlett Packard, IBM, NEC, and Silicon Image. The objective of the Digital Display Working Group is to address the industry's requirements for a digital connectivity specification for high-performance PCs and digital displays.

Plasma Display Monitor

The different states of matter generally found on Earth are solid, liquid, and gas. Sir William Crookes, an English physicist, identi-fied the existence of a fourth state of matter in 1879, which Dr. Irving Langmuir, an American chemist and physicist, called plasma. Energy is needed to strip electrons from atoms to make plasma. The energy can be of various origins such as thermal, electrical, or light (ultraviolet light or intense visible light from a laser).

Plasma can be accelerated and steered by electric and mag-netic fields, which allows it to be controlled and applied. In a plasma display monitor, light is created by phosphors that are excited by a plasma discharge between two flat panels of glass. The use of phosphors, as in CRTs, limits their useful life to 20,000 to 30,000 hours.

LCOS Monitor

Liquid Crystal on Silicon (LCOS) is relatively new technology that is a mixture of micro mirror and liquid crystal technologies. LCOS devices can be smaller and are easier to manufacture than conventional LCD displays and have higher resolution.

OLED Monitor

Organic light emitting diode (OLED) technology uses substances that emit red, green, blue, or white light. Without any other source of illumination, OLED materials present bright, clear video and images that are easy to see at almost any angle. When used as pixels in flat panel displays, OLEDs offer advantages over LCDs that need backlighting, including lower power consumption, greater viewing angle, lighter weight, and quicker response. The OLED screen appears unusually bright because of their uncommonly high contrast.

Touch Screen Technology

Because touch screen technology greatly simplifies the computer-human interface, input technologies are moving from hardware-based buttons, membranes, and keyboards to software-based models. This software-programmable approach affords numerous benefits for a user-friendly interaction. There are three basic types of touch screen technologies: capacitive, resistive, and surface wave. Each has its own strengths and weaknesses.

Capacitive touch screens use a glass overlay with a thin metallic coating over the surface of the display screen. Touching the overlay surface causes a capacitive coupling with the voltage field, drawing a minute amount of current to the point of contact. Capacitive touch screens are typically brighter and clearer than resistive screens, but not as bright or clear as surface acoustic wave.

Five-wire resistive overlay technology consists of a glass overlay with a thin metallic coating, over which a layer of polyester is placed. The polyester has a similar metallic coating on the interior surface. Tiny spacer dots of non-coated polyester prevent the two surfaces from contacting each other. A final hard coating is usually applied to the external surface of the polyester to reduce damage from sharp styli. A current is pulsed through the glass overlay along the x-axis and then the y-axis. When a finger or stylus presses the two layers together, the current is shunted and the control electronics determine the coordinates of the touch location, which are then transmitted to the host computer.

Surface acoustic wave is based on transmitting acoustic waves across the surface of a glass overlay placed over the display surface. A transducer mounted on the edge of the glass emits the acoustic wave, which travels on the surface of the glass overlay. When a stylus such as a finger comes into contact with the wave, it decreases the amplitude of the wave motion, absorbing part of the wave. This is detected by the control electronics and determines the touch location.

As these few elements of monitor technologies point out, the science behind video display devices can be somewhat complicated. Adequate research and investigation should be given to the selection of these devices just as is given to other components of the video system.

PRINTERS—WHAT YOU SEE IS WHAT YOU GET

The roles of printers in conjunction with security applications, from printed incident reports and identification cards to the replication of video images, require reliable products that produce the highest quality printed media. Computer printing moved through several stages of innovation, from the first daisy-wheel and dot matrix (or impact printers) to common use of the non-impact printers: ink-jet, laser, thermal-transfer. Now, the standard is digital printers.

The printer is especially important in access control badging applications where the exodus from conventional photographic-based ID badging toward digital imaging and printing has resulted in a real ability to integrate ID badging with CCTV quality ID badge. Today's digital printers allow for the ability to integrate information contained on the badge into the overall access control system. Personalized card templates are sent to a desktop card printer where personal identity information gathered during the capture process is printed or stored on the card in a single-pass operation. Photos, text, bar codes, and other graphics are printed on one or both sides of a card. Biometrics, cryptographic data, and other machine-readable information is loaded into smart cards, encoded onto magnetic stripes or printed as bar codes. Finally, the printer applies a polyester overlay to enhance card security and durability.

It is obvious that many details are involved in the choice of printer for the application. In order to make the best selection for your job and budget you need to determine the goal of your printer and proceed with careful research of products and companies.

11

Integrating Digital Video with Other Technologies

One of the most interesting aspects of the security industry is the multifaceted utilization of its products and services. It is comparable to the communications industry in its versatility of end users and uses. Security products and services are found in many areas—residential, commercial, public service, transportation, industrial, and military. Not only does the security industry supply a limitless market, it also combines with many cross markets to create efficiency and economy of products and services.

"Systems Integration" became a security industry buzz word in the late 1990s and post Y2K era. Technology justified the term by making it possible to interconnect, interface, and integrate subsystems of countless varieties, all of which resulted in a marked increase in security systems sales. Security system dealers and installers became more commonly known as systems integrators, and integrated security systems simplified both maintenance and operations, resulting in a reduced total cost of ownership.

Customers want integration for the advantages it provides, but barriers like custom and proprietary backbones of existing equipment have to be considered. Historically, manufacturers believed having a proprietary protocol protected them from competitive vendors, but today, the opposite is true. Customers are demanding open architecture and common protocols in order to reap the benefits of integration such as the cost savings incurred from streamlined business processes and increased efficiency.

The Security Industry Association (SIA) has identified the need to clarify systems integration and has created The Systems Integration Industry Group (SIIG), a group of security professionals who are tasked with defining integration and establishing methods and standards for the integration sector. The mission of SIIG is to create an environment where members of the Security Industry can gather to communicate the needs facing those who are active in the integration sector.

INTEGRATED VERSUS INTERFACED

The term integrated is often used loosely to describe the result when two or more systems are connected to work in conjunction with each other. Systems are often described as integrated when they should more accurately be described as interfaced. When a system is interfaced with another system, an event on one system can trigger an event on another system. For example, a door opening on an access control system could trigger a camera to pan, tilt, and zoom to achieve better coverage, or could change the record rate of the images from the appropriate camera. See Figure 11-1.

When a system is integrated, similar triggers have the same effect, but the integrated system goes a step further. For example, a card presented at an access control door may cause the appropriate camera to pan, tilt, and zoom for better coverage. It might then display a live image from that camera along with the badge holder's picture for verification. With the interfaced system, the video would be displayed on one monitor or workstation, while the access control data is displayed on another. With the integrated system, an operator could potentially deny access through the door if the person in the live image presenting the card does

Interfaced Systems

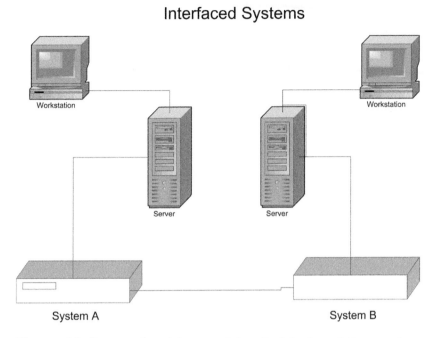

Figure 11-1 Interfaced Systems Must Be Monitored Separately

not match the image on file as the authorized badge holder. See Figure 11-2.

Thanks to advances in compression and telecommunications technologies, remote video can combine several security systems into one that is both competent and cost effective. The basic remote system is composed of CCTV cameras installed at locations where unauthorized intrusion, employee theft, or other criminal activities may occur. A video transmitter is integrated with the CCTV system that connects to a receiving site. This connection may be initiated by the sending or the receiving location, either manually or by automatic alarm triggers. In the case of an alarm trigger, strategically placed alarms will alert the receiver of security breaches and begin providing live video, audio, and in some cases specific data about the incident as it is occurring. An audio feature can allow a receiver to announce his or her presence and inform perpetrators that they are being observed and recorded.

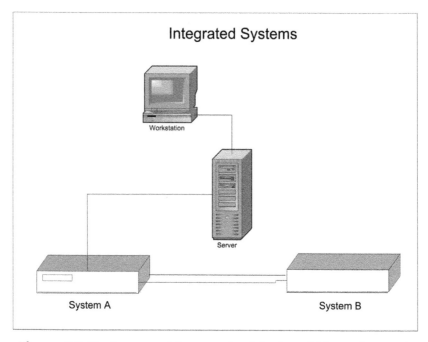

Figure 11-2   Integrated Systems Are Monitored Through a Single User Interface

One of the greatest advantages of integrating video is alarm verification. When an alarm is activated, the receiver can immediately view scenes of the alarm location, assess the information, and take appropriate actions to alleviate the situation. Unnecessary calls to law enforcement are virtually eliminated. Another distinct advantage of remote video is that information is stored, providing documentation of events.

BIOMETRICS

Biometrics is the science and technology of establishing the identity of an individual by measuring physiological or behavioral features. Because it can be easily incorporated into surveillance applications, facial recognition technology for identification and authentication is experiencing significant growth in both the public and private sectors.

According to the National Defense University in Washington, D.C., biometrics refers to the utilization of measurable physiological and/or behavioral characteristics to verify the identity of an individual. In an authentication system, the goal is to confirm whether the presented biometrics match the enrolled biometrics of the same user. Biometrics falls into two categories: physiological and behavioral. Common physiological biometrics authentication includes such things as face, eye (retina or iris), finger (fingertip, thumb, finger length or pattern), palm (print or topography), hand geometry, and wrist, vein, or thermal images. Behavioral biometrics includes behaviors such as voiceprints, handwritten signatures, and keystroke/signature dynamics.

These systems identify individuals by comparing known images to live images from a camera. This means that the camera system now becomes an integral part of the access control system, with the live images helping to determine whether access is granted or denied. By adding multiple cameras, it is then possible, in theory, to search a building for a specific person based upon the last known location. It is also possible to search crowds of people for specific individuals, such as those stored in terrorist or criminal databases.

When facial recognition is used for access control, the person requesting access usually must initiate a comparison, such as by presenting a card to a card reader. The facial recognition system then only has to do a "one-to-one" comparison, comparing the live image to the image on file for that card holder. This is also known as a verification test. When facial recognition is used to monitor crowds, there is no means of initiation and the system then is performing a "one-to-many" comparison. The live image of the person in question must be compared to the entire database of images to determine if that person is in the database.

A form of thermal imaging called a thermogram reads the facial heat pattern using an infrared camera. The identification process begins by capturing the multitude of differences in each human face. Every human thermal facial image is unique to an individual and remains consistent from birth through old age. Even identical twins do not share the same infrared image. The amount of heat emitted from an individual's face depends on nine factors, including the location of major blood vessels, the skeletal system thickness, and the amount of tissue, muscle, and fat in the area. Presently, the most accurate biometric besides thermal is an iris or retina scanner, which is significantly more expensive than face, finger, or palm recognition systems. It is also harder to fool.

ACCESS CONTROL

To understand the advantages of incorporating video with access control, it is important to first understand the purpose of the access control system. Access control is used primarily to allow or deny access to individuals through controlled and monitored points within a building. Typically, employees or others who are meant to have access to certain rooms, areas, or buildings are issued cards that must be presented at card reader locations to obtain entry. Typically, this card is used as an identification badge; therefore it contains employee data and often a photograph of the intended cardholder. The card also carries information about any restrictions that may apply, such as when and where entry is authorized.

Access control card systems range from inexpensive, stand alone systems where the microprocessor is located in the door without recording capabilities to more expensive systems which link multiple doors to a central computer. When a card is inserted into the latter type of access control unit, information from the card is sent to the computer where validation and recording functions take place. The control of access is performed by a card reader. Choices of card readers generally include proximity, weigand, magnetic, or bar code.

Proximity readers, as the name implies, depend upon the card's proximity to the reader. The most popular of these readers work when a card is presented within approximately five inches from the reader. There are readers that will work from a distance of three feet. The main advantage to using proximity is the ease of use—the user need not stop and insert the card into the reader but merely make sure that the card is within the prescribed ranged of proximity. In some cases, the card itself may even remain in a purse or wallet while activating the reader.

Weigand card technology consists of a series of specially treated wires, which are embedded in each card. These treated wires possess unique magnetic properties. When the card passes through the reader, a sensing coil picks up this unique signature and transmits it back to the controller.

Magnetic cards are encoded with information that is read by swiping the magnetic stripe through an appropriate card reader that senses the code. The process used to make magnetic cards is relatively simple, consisting of a stripe, which is a coating of iron oxide or other substance that can be magnetized and demagnetized. Some magnetic stripes require more coercivity than others. Coercivity is the strength of a magnetic field required to record or change data on the magnetic strip. Everyday magnets can erase a low-coercivity magnetic stripe; those with high coercivity are virtually non-erasable.

Bar codes are graphical representations of information encoded within a series of bars and spaces. All bar codes have certain bar code patterns which tell the reading device when to start reading the bar code.

The weak link in a standard access control system is often the lack of verification of who is presenting the card at the reader. If a card is lost or stolen, the card reader will still function when the card is presented until it is disabled in the database. Biometric devices can help to eliminate the possibility of using a stolen card, but they cannot always verify that an employee is not entering under duress. Some devices will have the possibility of using a different body part if under duress, such as using the right eye instead of the left on an iris recognition reader. If the employee forgets, however, it is possible to have a false duress read or a missed duress read.

Adding video coverage at access control points can enhance the system in several ways, depending on how the system is monitored. It is most advantageous when the access control system is monitored in real time by an active protective force. When this is the case, an operator can verify that the card being read is in the possession of the rightful owner and that the cardholder is not under duress.

With a system that is integrated in this manner, an active card read will automatically display the proper camera on the monitor that shows the door that is being accessed. In addition, the badge photo that is in the database can be displayed directly next to the live image, allowing the operator a comparison of the person at the door and the person authorized to use the card presented.

Video integration can also display live camera views for the operator in other situations. An attempted entry with an invalid card or a card that is presented outside of the authorized access times can cause the appropriate camera to be brought up, allowing for a live assessment. With an integrated system, it is also possible to search for specific things, such as an individual cardholder.

For example, if an employee is suspected of taking something such as a laptop, the investigator can search for the associated employee to see which doors he or she accessed. The investigator will then have a reduced amount of video to review to see if the employee can be seen leaving with the item. If the access control system requires personnel to use a card reader to exit (read in/ read out or anti-pass back configurations), the investigator can go directly to video of the specific time that the employee exited.

Many central stations now have the ability to view live video when an alarm occurs, thus allowing them to make an informed decision prior to dispatching first responders. If the intrusion detection system sends an alarm to the central station indicating that a specific entrance has been breached, the operator can access live video to visually check the situation. If all appears normal, a review of the time immediately prior to the alarm can be done to see what may have caused the alarm to be triggered. If the video still shows nothing unusual, the operator may determine that a false or nuisance alarm has occurred and choose not to dispatch authorities. Usually, in this case, an owner or designated contact is summoned to take appropriate actions.

PERIMETER PROTECTION

The level of protection provided for the protection of a building or area is determined by the level of risk from intrusion and is often comprised of several different, complimentary layers of protection. Perimeter protection can include any combination of things

Figure 11-3 Mobile surveillance tower from P.I.C.S (Portable Intellegence Collection System)

like bollards, security fencing, barriers, turnstiles, doors, bars, grilles, motion detectors, PIR, open ground electronic protection, or radio frequency intruder detection. The addition of video surveillance cameras at the perimeter can make a significant contribution towards tightening the whole security system. See Figure 11-3.

Video technology is commonly used to enhance perimeter security at correctional facilities. Video technology not only improves security but also replaces the need to man gun towers and allows for a reduction in armed perimeter patrols. Electronics have, in many cases, entirely eliminated the need for towers and the construction costs associated with them. The strategy has been to strengthen the entire perimeter with double fences bristling with electronics and have one or two patrol vehicles (rovers) constantly circling the facility with armed officers. As a result, staff previously assigned to these posts could be shifted to other, more critical areas.

External active infrared detection has been in use for perimeter protection since the late 1920s. These detectors utilize active infrared beams to detect unauthorized entrance or movements through an invisible barrier. An active infrared beam, also called a photoelectric beam, is a sensor that transmits a focused infrared beam, which is received by a photocell and responds to an interruption of the beam. Active infrared detection is susceptible to the false alarm.

Video surveillance installed at many sites using active or passive infrared detection can be effective in some cases, but verification of alarms at external sites especially can be hindered by weather and light conditions. Unless all of the cameras are equipped with thermal imaging devices, some scenes will necessarily be missed or unidentifiable. Another difficulty is pinpointing the exact location of an alarm. With infrared beams capable of reaching in excess of 200 meters, the result is a potential intruder located anywhere within a 200 meter zone.

12

More Digital Video Applications

Law enforcement facilities and correctional institutes are primary applications for digital video surveillance systems. Video is used for a variety of purposes in these facilities including security, evidence of brutality against prisoners, videoconferencing, and even for the provision of medical care via telemedicine technology. Surveillance levels depend upon the security level of a facility. These levels are minimum, medium, maximum, and super max. The higher the level of security, the higher the number of cameras installed. A super max facility has virtually no area outside of CCTV view.

Prison visitors are not exempt from the auspices of video technology. CCTV is often used in prison visiting rooms, for observing treatment programs, and for auditing mandatory drug testing of prisoners. This helps to minimize the time required to clear visitors into and out of correctional facilities as well as reduce the number of corrections officers involved in visitor processing.

When a prison utilizes this type of system, a visitor must pass an authentication process before being allowed to visit a resident.

During this process, the officer may be viewing a live video display on the PC screen from a CCTV camera. The resident's information is displayed, logged on a printed report, and saved in the central database. The visitor comes to the secured door and keys in their visitor identification number. A valid number will bring up the visitor's image on the correction officer's screen. The screen also displays a list of approved residents for this visitor. At this time, an indication is given if visitation privileges have been revoked. The live video can be compared to the database image displayed on the screen where remote operation is required. The visitor states which resident or residents they wish to visit and access is either confirmed or denied.

An important aspect of video technology is its impartiality. Video cannot take sides; it can only display events as they actually occur. For this reason, video is often the advocate of the victim.

Numerous opportunities are available for using this technology in a correctional facility including employee training, business meetings, court hearings, and parole or deportation hearings. Videoconferencing and telemedicine technologies reduce the need to transport dangerous prisoners. Telemedicine programs offer significant safety, security, and cost advantages to correctional facilities while being able to provide the services of specialists not readily available to incarcerated individuals.

Public safety personnel around the nation are starting to use basic technology tools such as laptops, PDAs, and Automated External Defibrillators (AEDs). In 2004, Washington, D.C. launched the nation's first broadband data network for emergency crews, an important step toward arming rescuers with the latest communication technology. High-speed wireless networks allow emergency room doctors to see live video of a patient still in the ambulance or police helicopters to stream live video from the air to patrol cars on the ground. The technology enables all rescuers to talk directly to each other.

Telemedicine

Telemedicine has been defined as the use of telecommunications to provide medical information and services. It may be as simple as two health professionals discussing a case over the telephone or as sophisticated as using satellite technology to broadcast a consultation between providers at facilities in two countries using videoconferencing equipment. The first is used daily by most health professionals, while the latter is used by the military, some large medical centers, and increasingly by correctional facilities.

The University of Texas Medical Branch at Galveston was one of the original programs to begin providing services to inmates and sees over 400 patients per month. The foundation of the UTMB telemedicine network is a scalable, ISDN network operating over leased T1 lines. Once the technology was in place and real-world applications identified, the rollout began. One application linked 12 remote sites to UTMB to provide medical care for special-needs children in areas where medical technology and expertise were not readily available. The telemedicine solution included a virtual exam room with a video interface designed to be simple enough for medical personnel to operate, so that the bulk of their time could be spent treating patients, not manipulating video equipment.

Major specialties using the network are neurology, psychiatry, orthopedics, dermatology, and cardiology. The feedback from both patients and physicians has been positive, with access to specialty care and saved travel time cited as the most important benefits of the encounters. Using a variety of specialized patient cameras, comprehensive patient examinations can be performed, including diagnostic cardiac echo cardiology and ultrasound imaging. High-definition monitors allow the patient and the physician to interact as if they were in the same room. With the primary care physician and the specialist both involved in a medical consultation, pertinent history can be discussed and interventional therapies agreed upon.

For correctional facility managers, telemedicine may offer a means of providing appropriate health care evaluation without compromising security, reducing costs associated with transport and protection, and gaining access to physician specialists and resources unavailable within the prison medical system. Between September 1996 and December 1996, a leased telemedicine network was installed to serve four federal prisons to gather information on the effectiveness of this technology. One suite, located inside the penitentiary, served inmates at both the United States Penitentiary and the Federal Correctional Institution in Allenwood, Pennsylvania, another served inmates at the United States Penitentiary in Lewisburg, Pennsylvania, and a third served inmates at the Federal Medical Center (a prison health care facility) in Lexington, Kentucky. All of these sites were networked for telemedicine with the Department of Veterans Affairs Medical Center, also in Lexington. The VA and Federal Medical Centers in Lexington served as the hubs in this network, providing specialist physicians and other health care practitioners for remote (telemedical) consultations with prisoners in the three Pennsylvania prisons.

The purpose was to test the feasibility of remote telemedical consultations in prisons and to estimate the financial impacts of implementing telemedicine in other prison systems. One of the largest for-profit government and business consulting and research firms in the country, Abt Associates Inc., was contracted to evaluate the demonstration and estimate the costs and savings associated with the use of telemedicine in these selected prisons. During the demonstration, a fifth mode of care—remote encounters with specialists via telemedicine—was added to determine whether the prisons could use telemedicine to overcome local problems in accessing needed specialists and improve security by averting travel outside the prison walls. The demonstration was also designed to supply data on costs and utilization to support a decision about whether and where to implement telemedicine in other prisons.

In a press release issued in mid-1999, results of a report from Abt Associates (Cambridge, MA) highlighted the potential for telemedicine to reduce health care costs in prisons, based on data gathered in the prison telemedicine demonstration. Specifically, use of telemedicine systems instead of traditional forms of care (prison staff, in-person clinics, or other health care facilities) was estimated to save approximately $102.00 per specialist encounter. Other advantages were quicker access to care (reduced waiting between referrals and actual consultations) and use of physicians from outside communities who offer more competitive pricing for their services.

A telemedicine program at Louisiana State Penitentiary (LSP) is an outgrowth of the Louisiana State University (LSU) Medical

Center's telemedicine initiative that began in 1995. Before the telemedicine program, approximately 3,000 inmates from LSP were transported to the secondary and tertiary hospitals for medical-related reasons during a six month period.

The goals of this project were to reduce the number of inmate transports from LSP to the secondary and tertiary health care service centers, reinforce the security parameters and performance objectives of the Department of Public Safety and Corrections, and reduce the physical presence of inmates in the general civilian population served by hospital-based clinics.

Videoconferencing

Digital systems are used to communicate with federal courts to conduct pre-trial, civil, and mental competency hearings when it is not desirable to transport a particular inmate to court. Prison staff is encouraged to use videoconferencing as a means to reduce travel costs and reduce the risks involved in transporting prisoners.

Arizona's popular Sheriff Joe Arpaio made international news by transmitting live video from the jail onto the Internet for public viewing. The site provides real life transmissions from the Maricopa County Sheriff's Office Madison Street Jail. Maricopa County is the fourth largest jail system in the world. Housing over 1500 prisoners on average, the Madison Street Jail books an average of 300 suspects a day. The Office, headed by Sheriff Joe Arpaio, is known throughout the world for its tough stance on how inmates are incarcerated and overseen. Sheriff Arpaio is convinced that using video surveillance and the World Wide Web will deter crime. It is his hope that the only visit anyone makes to his jail is the virtual visit provided by the jail cam site.

As with any new procedures or technologies introduced into use at correctional facilities, video conferencing must pass certain criteria. The American Society for Testing and Materials (ASTM) Committee F33 on Detention and Correctional Facilities meets four times a year in conjunction with the American Jail Association (AJA) and American Corrections Association (ACA) Conferences to construct guidelines. The Operational Controls Subcommittee,

F33.06, has completed the revised "Standard Guide for the Selection of Operational Security Control Systems", ASTM F1465-03, and has started a new work item to develop a guide standard for the selection of digital video recorders (DVRs). In the future, this group plans to develop a standard for "Standard Terminology for Security Control Systems" and a selection guide for "Video Arraignment and Video Visitation Equipment".

Law Enforcement and Video   The Law Enforcement & Emergency Services Video Association (LEVA) is dedicated to serving the unique needs of law enforcement and emergency services professionals who use video. Whether it's video for production, training, surveillance, crime scenes or documentation, through its members, LEVA has established itself as the premiere source for information, quality training, and networking. Chartered in 1989, as a volunteer, nonprofit organization, LEVA serves videographers and audio/visual specialists from local, state, and federal law enforcement, fire, emergency medical, rescue, and other related public safety agencies throughout the world.

Although LEVA does not endorse any particular manufacturer or company product, its members are very knowledgeable about video equipment and are consulted by their employers and other public safety agencies for recommendations of potential purchases of video equipment. Areas of knowledge and expertise include in-car video systems, surveillance video equipment, crime scene and documentation equipment, training, and also multimedia and production equipment.

Large amounts of people, traffic, and excitement make significant public events a challenge for law enforcement. Not to be left behind the digital movement, Louisville, Kentucky businesses have begun converting their previously analog systems to the new digital products. The city sets up even more cameras for the famous Kentucky Derby festivals and events, which law enforcement use to monitor crowds and observe traffic patterns.

Getting three-quarters of a million people in and out of the venue is a monumental task that requires patience and extensive planning. The Video Forensics Analysis Unit of the Louisville

Metro Police Department was instrumental in planning and implementing the digital video surveillance aspect of security for the past few years' events. With digital video in place, law enforcement can view camera images via microwave, giving them the ability to direct support to specific locations when needed.

The Civil War marked a number of important technological advances that changed the methods used to gather and communicate intelligence. Photography was used for the first time. Aerial photography was also carried out, using hot air balloons. Additionally, telegraphy was used for the first time though messages were often intercepted and deciphered. By the time World War I came along, technology had advanced to include signals intelligence that gained greater importance than in any other war. Telegraph and radio messages, in Morse code, were soon vital to the conduct of war.

Covert Video

Covert video is accomplished fairly simply as almost any normally occurring piece of home or office equipment can hide a video camera, including lamps, books, smoke detectors, clocks, and even stereo components. 2.4 GHz transmitters do better indoors because the signal frequency is much smaller in width and can move through walls, in between the studs, and through rebar (which is in concrete or brick walls). The only limitation is that they cannot penetrate solid metal walls; the signal frequency will bounce off or reflect away. This makes it possible to place a covert camera or radio (2.4 GHz) in a room with the receiver, antenna, monitor, and recorder up to 500 feet away. Some types of covert equipment can be hidden on a person, but generally speaking, microwave signal frequencies (2.4 GHz) should not be worn on the body.

POLITICAL EVENTS

Nowhere are the efforts against crime and the use of technological tools, including video, more prevalent than in the protection of our nation's political figures. Among its uses for surveillance, live broadcast, and documentation for the public, video is the star of political events. During an election year, there are three major political events where security, technology, and broadcast video are tested to their limits:

- Political Party Conventions
- President and Vice President Candidate Debates
- Presidential Inauguration

While a specific discussion involving the particular elements of security at these events cannot take place, a general examination of the significant role of video can occur. When then Texas Governor George W. Bush received the Republican Party's nomination for President of the United States, he was under a protective umbrella of five high-tech command centers designed to prevent and respond to terrorist attacks or natural disasters. The Federal Emergency Management Agency, working with the Secret Service, the Environmental Protection Agency, and other federal, state, and local security agencies, established primary command centers in the Philadelphia region during the convention. Officials estimated the total number of people in and around the First Union Center in Philadelphia at more than 35,000. Given such a large, compact crowd, planners placed a premium on incident reporting and timely response coordination, according to the Federal Response Plan.

Security officials began to set up the high-tech monitoring effort three days before the convention. One of the primary sites established was the Secret Service's Multi-Agency Communication Center (MACC). The EPA and Secret Service officials staffed the MACC around-the-clock during the convention.

13

From VTRs to VCRs, DVRs, and NVRs

After the ability to create moving images was achieved, the next challenge was to record the moving images. One of the first men to attempt a type of electronic recording of information was an American mechanical engineer named Oberlin Smith, who came up with the idea of recording electrical signals produced by the telephone onto a steel wire. Although Smith never actually pursued his vision, he publicized his ideas in a journal called *Electronic World*. Later, Valdemar Poulsen, a Danish telephone engineer and inventor, patented the first apparatus for magnetic sound recording and reproduction. It recorded, on a wire, the varying magnetic fields produced by a sound. The earliest known attempted use of magnetic recording to store images was in the late 1920s, by Boris Ritcheouluff of London. Ritcheouluff designed a picture recorder based on Poulsen's machine, developed in Denmark many years before. Many more video recording concepts followed.

VIDEO TAPE RECORDERS—VTRS

The first practical videotape recorder (VTR) was developed and sold by Ampex Corporation in 1951. The Ampex VTR captured live images from television cameras by converting the information into electrical impulses and saving it onto magnetic tape. VTRs were similar to reel-to-reel audio tapes, with large spools of multi-track magnetic tapes measuring 1/2″ to 2″ wide and averaging 7,000 feet in length. The company demonstrated its Ampex-Mark IV in April of 1956 at the National Association of Radio and Television Broadcasters (NAB) convention. See Figure 13-1. Ampex sold the first VTR for $50,000 in 1956.

The lack of interchangeability among the very early VTRs posed a serious problem. The same head assembly used to record a program had to be used for playback, meaning that the recording machine head assembly had to be shipped with the tape of evidence in order to view the recorded contents. This problem still exists in some systems today because many manufacturers use proprietary compression codecs for their digital recorders.

Figure 13-1   Ampex-Mark IV VTR. Courtesy of Department of Special Collections and University Archives, Stanford University Libraries.

VIDEO CASSETTE RECORDERS—VCRS

Magnetic tape recording came into play near the end of World War II. Though the size of the tape, the speed at which the tape passed the recording heads, and the way video is written to the magnetic tape have changed over the years, the basic principles have remained the same.

The first Betamax VCRs were sold by Sony in 1971. In 1976 Panasonic and JVC introduced its competitor, the Video Home System (VHS). Originally standard VHS type video cassette recorders were used for CCTV applications. A VCR is a device that can record images from a video camera onto magnetic tape; it can also play pre-recorded tapes. It is helpful to understand the mechanics of VCR recorders to also understand the shortcomings. Videotape is a plastic ribbon impregnated with a magnetizable metal powder. Before recording, the particles are oriented randomly. During recording, the video heads create a magnetism that orients the particles in certain directions converting video signals into magnetic patterns on the tape. When the tape is played back, video heads again pass over the magnetic powder and sense the magnetic vibrations and convert these vibrations back into a video signal.

Video signals consist of millions of electrical vibrations each second. Each vibration represents a tiny piece of your picture. Videotape is a ribbon of Mylar with billions of tiny magnets glued to it with a sophisticated kind of glue called a binder. See Figure 13-2.

The original magnet material was iron oxide, and so the side of the tape with magnetic material on it is called the oxide side. Notice the irregular surface of the oxide coating. When the tape is being manufactured the mixture of magnetic material and binder (glue), called slurry, is liquid. The slurry is spread over the Mylar and allowed to cure or dry. When the slurry cures, some of the magnetic material protrudes from the surface of the oxide side.

When the video is "written" to the magnetic tape, the video heads are in intimate contact with tape. In fact, the heads actually protrude into the surface of the tape, causing a "canoe" in the Mylar surface. As a result of this contact, the video heads

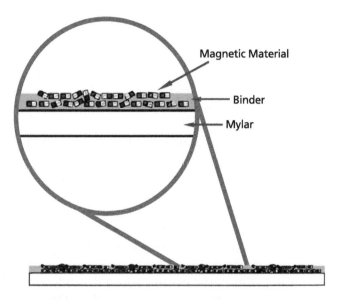

Figure 13-2 Construction of Video Tape

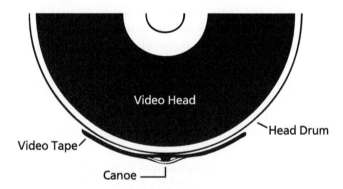

Figure 13-3 Video Head Wear

experience wear from the friction of the tape rubbing the heads. See Figure 13-3.

In addition, there is some oxide rubbed off from the tape. Video headwear is especially high when most of the tape is brand new because of the irregular surface of the oxide layer. As the heads "burnish" new tape during recording, the oxide layer surface is more abrasive than slightly used tape. The surface irregularities are literally "sanded" off the oxide surface. As a result, the oxide layer develops a very smooth surface. After many hours of use, the worn out heads have to be replaced.

Debris from the oxide and head wear collects around all of the guides. The result is that the tape transport mechanism must be cleaned from time to time, or the oxide debris will build up and cause the edge of the tape to be damaged. If the edge of a tape becomes sufficiently damaged, the tape will no longer yield a good quality of recording and playback. Debris also can clog one or more of the video heads. This results in loss of recording and large snowy areas in the picture on playback.

The transport mechanism is complex for a thin, half-inch ribbon of Mylar, as Figure 13-4 illustrates. Notice that from the supply reel to the take-up reel, the tape must go through three 180 degree turns. This complex tape path results in a fair amount of stress on the Mylar that, in turn, results in tape edge damage and overall quality degradation.

Industrial video recorders differ from consumer recorders in several ways. For example, they usually operate 24 hours a day

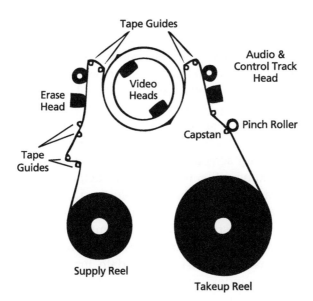

Figure 13-4 Transport Mechanism

and seven days a week in time lapse mode, which allows a recording of extended periods on video cassettes. High grade video cassettes are needed to avoid damaging video recording heads. Industrial grade tapes should be replaced after multiple uses.

DIGITAL VIDEO RECORDERS—DVRS

The biggest selling feature for digital surveillance to date has been the switch from the video cassette recorder storage to digital storage. The combination of affordable image compression technology and large capacity hard disks made the development of digital video recorders feasible. Digital video storage or digital video recorders (DVRs) are a practical replacement for analog VCRs because of the elimination of problems such as poor image quality from the reuse of tapes, worn out heads, scratches, and stretching from searching back and forth for a specific scene. Wear and tear aside, no matter how many guidelines are set up for the management of conventional video tape, one of its biggest downfalls is the simple action of having to place a tape into the VCR.

A digital video recorder is a stand-alone unit capable of saving images to a hard disk. DVRs look similar to a standard VCR in some ways, but that's where the similarity ends. Because digital systems are not mechanical like VCRs, factors such as frame speed and video quality are software adjustable. Unlike the VCR, a digital video recording device provides clear, sharp images every time it is played. There are no tapes to store and material does not deteriorate over time. A digital system allows for auditing of activity through monitor screen menus and for images to be retrieved as easily as opening a file, using criteria such as date, time, location, or camera number. Whatever role the DVR plays, its very existence declares a system to be digital even though the camera, transmission, and display technologies may be analog. Digital video storage allows particular images to be retrieved as easily as opening a file based on criteria such as date, time, location, camera number, special index numbers, etc. Digital video storage eliminates the need to store hundreds of space consuming VCR tapes, and archived material does not deteriorate with time.

The history of the DVR involved a wide variety of technologies and manufacturers. Costs and availability varied greatly as well. One of the first companies to deliver a successful product was Dedicated Micros of Manchester, England. The Dedicated Micros DVST (Digital Video Storage and Transmission) competed closely with a Sensormatic remote transmission product.

In the 1990s, motion detection was added to the management software of the camera or recorder, giving us the first hint of "intelligence" in the systems. There are now literally hundreds of DVR manufacturers with a wide variety of products and features, many specializing in solutions by size, location, lighting conditions, or number of cameras. DVRs are usually scalable and upgradeable utilizing specific software. They typically have video capture circuits or cards that can process 60, 120, 240, and 480 frames per second. These numbers represent the total number of frames per second that can be accommodated for all of the cameras or channels per system. For example, the 120 frames per second DVR with 16 cameras has an approximate frame rate of 7.5 frames per second. This means that each camera can be converted at 120/16 or about 7.5 frames per second.

NVR—NETWORK VIDEO RECORDING

New video storage systems work with network attached cameras. This new technology is very flexible and provides excellent features that allow you to create a complete video surveillance system. Network Video Recording is a digital video recording solution that works over a TCP/IP network. IP addressable network cameras and/or video servers transmit images over a LAN, WAN, or across the Internet. A NVR automatically receives data from any IP cameras on a network and store it locally or on remote storage media. The data can be any combination of video and audio with hundreds of streams stored on a single server.

The frequently used term client/server describes the relationship between two computer programs or, in the case of networked video, the NVR and the remote network cameras. The client makes a service request and the server fulfills the request.

In a network, the client/server model provides a convenient way to interconnect programs that are distributed efficiently across different locations. When speaking of the Internet, for example, a web browser is the client program making requests from a web server connected through the Internet.

Using IP surveillance, the recording function is performed by a network video recorder—a standard PC Server loaded with video recording software. The NVR accesses the data streams of the remote network cameras and video servers and stores them on hard disk. Even though this technology is backward compatible with all legacy analog CCTV equipment, more and more network cameras that can interface directly to IT-style Ethernet networks are being used today.

NVR technology is cost efficient and with IT-concepts such as Storage Area Networks (SAN) or Network Attached Storage (NAS), capacity can reach the terabyte range.

DVS—DIGITAL VIDEO SYSTEM

Just to make life interesting, some people refer to the digital video recording system as a DVS for Digital Video System. Just like the DVR, the DVS is used for capturing, recording, displaying, archiving, and retrieval of video images, usually in a PC-oriented environment.

14

Central Station Monitoring and Video

Retaining the services of a video central station to remotely monitor video images has the advantage of delegating the work of product selection, installation, maintenance, communications coordination, training, and monitoring to one agent who can be located outside the monitored facilities. The basic remote digital video system involves CCTV cameras installed at remote locations where intrusion, criminal activity, or employee pilferage may occur. A video transmitter is integrated with the CCTV system, which has the ability to dial-up the remote video monitoring center. This dial-up function will be activated either by the video monitoring center or by an alarm triggered at the site.

In the case of an alarm trigger, strategically placed alarms will alert the video monitoring center of security breaches either by unscheduled openings, "panic buttons" activated by an employee, or movement in sensitive locations. When an alarm is triggered the system will automatically dial up the video monitoring center and begin providing live video of the scene.

Alarm Verification

There are almost limitless opportunities for applying remote video services to improve profitability and enhance business performance. Alarm verification is a key use of remote monitoring operations. Visual alarm verification goes right to the core of solving the false alarm issues. Unnecessary calls for police service due to false alarms have grown into an enormous problem both for law enforcement and security providers. It has been estimated that police are responding to somewhere between seven million to fifteen million false alarms every year. In light of these continually increasing numbers many police agencies have had to assess fines to supplement rising costs, which are a direct result of the false alarm dilemma. False alarm fines can range from $20 to $250 or more. If false alarm fines are affecting a company's profitability, remote video alarm verification can help dramatically reduce these costs.

Not only can an alarm be verified but also a detailed description of events is available for the responding security patrol or law enforcement personnel. Facts about approximate age, demeanor, and number of persons involved, and if they have weapons can be made available before a response team arrives at the location or even sent directly to response vehicles if they are equipped to receive video images. This vital information can augment the safe and successful resolution of many incidents.

When a video monitoring station receives an alarm, a trained specialist can immediately view the alarm location to determine the status of the alarm. If an alarm is determined to be false, a remote system can automatically record information pertaining to time, date, and location. The remote monitoring location can record the cause of alarm for future reporting. In the case of an actual unauthorized intrusion, the specialist can document all information and inform the appropriate authorities immediately.

REMOTE INTERACTIVE MONITORING

Remote interactive video not only provides the communication of video to outside locations but simultaneous two-way audio as well. This technology allows communication between the remote site and trained intervention specialists who can clarify, deter, or diffuse transpiring events.

With the availability of simultaneous audio and video transmission, the intervention specialist can influence the outcome of events either by announcing his or her presence, issuing specific voice commands, or contacting local law enforcement with details of the situation. A pre and post alarm feature is usually available that insures that the stored video clearly displays the events leading up to and immediately following an alarm.

Remote interactive digital video solutions permit the manipulation of objects from the visual command center. Pan, tilt, and zoom features on cameras can be controlled. Motorized gates, electric door locks, lights, and even environmental controls can be remotely actuated as well. On site personnel may also call up a monitoring center to report suspicious behavior or other concerns and ask for video monitoring and potentially intervention. Panic buttons can also be provided which, when activated, will alert the center of a possible incident if the client cannot overtly call for help.

Video Guard Tours

At random intervals, a monitoring station can connect to the remote site and conduct a visual assessment of the facility. These tours can be announced to the facility, letting the employees know that they are not alone and increasing their feeling of safety. The tours also can be conducted as unannounced. This option lets employees know the facility is being viewed but they don't know when or how often. This has proven over time to have a significant impact on reducing internal shrinkage. Many users actually utilize both types of guard tours to maximize the capability of the technology.

Video guard tours can provide reports containing important information about operations. The following are examples of possible items that might be reported:

- Employee rude to customer
- Employees not following security procedures
- Team members not wearing uniforms
- Clerk working out of open cash drawer
- Open safe
- Unauthorized visitors or phone calls
- Pilfering
- Unsuitable work performance

Video guard tours increase the overall feeling of comfort and safety for employees and customers alike because they know they can quickly get help when they need it. They can also increase productivity without the necessity of constant upper management supervision.

Open/Close Escort Service

Remote monitoring services may include an escort service for employees at the opening and closing of a store, financial institute, or other facility. This service involves a live look-in during the opening or closing (by an authorized employee) to ensure safety and provide verification of events related to the procedures involved with opening and closing. This service gives the employee the ability to signal the monitoring station before they leave the building to take a deposit to the bank, take the trash to the dumpster, or leave for the night. They would then follow the employee with exterior cameras to ensure his or her safety.

Controlling Inventory Shrinkage

Each time someone takes items from a location without paying, the bottom line suffers. It may be a no sale transaction to an acquaintance or merchandise carried out the back door. Rather than review register tapes and inventory records, a video record of the incident not only identifies the problem but also provides indisputable proof of the event. The ability to monitor doors and loading docks and to vocally address each person in the site by their physical description will help to make the environment less inviting to shoplifting, buddy discounts, and back door transactions.

Remote Site Management

Remote facilities such as electric sub-stations or powerhouses are prime targets for vandals and thieves. Most of these stations are scattered throughout a particular region, making electronic monitoring without video capability or an on-site guard a time consuming and expensive proposition. Remote digital video and audio services can eliminate much of the travel and guard expense, giving managers the ability to out-source many of the operations needed to monitor their facilities.

The option of transmitting live video directly to the home or office of management is also available. When an alarm is activated and a premises needs inspection, a specialist can be on the scene in seconds to see what is taking place and to record the events. A predetermined course of action, which has been specified appropriate under those circumstances, can be initiated. It may be any combination of direct audio intervention, calling the authorities, or calling management.

Retail facilities benefit greatly from remote video monitoring for many reasons. Store transactions can be observed remotely if necessary as well as recorded and archived for future review and dissemination. This information can be vital when following up on situations of credit card fraud, checks of insufficient funds, or the presentation of false identification.

Loss prevention for retail establishments has evolved through various deterrents such as security guards, observation mirrors (that allow staff to see throughout the store), and closed-circuit television surveillance systems. These techniques were among the earliest tools used to combat shoplifting. Some larger stores still use a combination of security guards, CCTV systems, and even mirrors, but the advent of digital video technologies has made a great impact on how CCTV can be used to increase efficiency and decrease manpower.

The presence of unruly crowds, vagrants, and even graffiti can deter customers from patronizing an establishment. Remote interactive monitoring of facilities can alleviate these situations while promoting the image of a safe and secure environment for customers.

Today's retail management has the ongoing task of improving the customer experience while at the same time keeping losses from liability and fraud under control. Loss in the retail world is not subjective. The small boutique and the nationwide chains are both equally susceptible to losses from shrink, fraud, and shoplifting. With digital video in place, store transactions can be observed remotely if necessary, as well as recorded and archived for future review and dissemination. This information can be vital when following up on situations of credit card fraud, checks for insufficient funds, or the presentation of false identification.

The knowledge that remote monitoring is in place can be extremely beneficial to the peace of mind of employees, customers, and facility managers alike. With crime rates escalating, customers are known to patronize facilities that offer a reasonable expectation of safety. Not only is the emotional well being of staff and clientele important, but the decrease in liability incurred by facility owners emphasizes return on investment.

15

More Digital Video Applications

Security products and services are found in markets including residential, commercial, public service, transportation, industrial, military, etc. Not only does the security industry supply a limitless market, it also combines with many cross markets to create efficiency and economy of products and services. These are some examples of ways in which digital video is being utilized today.

KEEPING WATCH OVER PRODUCTION

Weyerhaeuser, an international company that offers a full range of pulp and paper products, provides an excellent example of how the provision of safety and security services and products consistently relies upon a blending of ingenuity and teamwork. By combining CCTV technology with computer network capabilities, they have utilized the full versatility of digital video in one of the housing materials manufacturing facilities. During the last 30 years, manufacturing plants have been automating their production facilities at an increasing rate, augmenting the need for monitoring systems that effectively track possible break downs and bottlenecks along the actual production line.

Most manufacturing plants and warehouse facilities depended on sensors and video cameras placed in many different locations to ensure proper functionality along manufacturing lines. At Weyerhaeuser, cameras are used by staff charged with engineering control systems in the various manufacturing facilities. The system allows them to log onto the Internet and observe the production systems at any time, from any location.

Weyerhaeuser discovered some additional benefits to the technology. The facility makes Oriented Strand Board (OSB), which is a wood panel material manufactured in 12 by 24 foot pieces weighing 500 pounds each. A jam in the OSB production system is not only costly in production time lost but also is hard to clean up. When installing and testing their new digital system, a potential jam in one of the production systems was observed over the Internet. Because of this observation, floor technicians were immediately contacted and consequently able to prevent the jam. According to Weyerhaeuser, a production jam that stops the manufacturing line can cost the company several hundred dollars a minute. Even the prevention of one jam per month can save the company thousands of dollars per year in time that was not lost.

SECURING A WAR HERO

The Battleship Missouri, nicknamed the "Mighty Mo", is located on Pearl Harbor's Battleship Row and opened as a floating museum on January 29, 1999. The 887-foot, 45,000-ton USS Missouri served in three wars—World War II, Korea, and Desert Storm—over a five-decade span. It is best known for being the site of Japan's surrender to the Allied Forces on September 2, 1945, ending World War II. Today, the "Mighty Mo" is berthed approximately 300 yards from the USS Arizona Memorial, and the two are memorials symbolizing the beginning and end of America's involvement in the world's deadliest war.

The USS Missouri Memorial Association is a private Hawaii-based 501(c) (3) non-profit organization designated by the U.S.

Navy as caretaker of the Battleship Missouri Memorial. The association was instrumental in the decision and planning processes that involved a state-of-the-art web cam system being installed on board the Missouri. The system allows the Association to bring the Battleship Missouri to the rest of world. Once complete, the system will provide a comprehensive virtual experience for physically challenged visitors through a new Visitor Alternative Media Center, allow far-away family members of U.S. serviceman to witness their reenlistment ceremonies held on the ship, and offer tours on the Internet for classrooms across the globe. In addition, staff can monitor daily activities and provide direction and assistance when and where needed.

REMOTELY MONITORING NUCLEAR MATERIALS

The U.S. Department of Energy, Aquila Technologies, Los Alamos National Laboratory, and Sandia National Laboratories launched collaborative efforts to create a Non-Proliferation Network Systems Integration and Test (NN-SITE) facility. Utilizing Aquila Technologies Group's Gemini system, the facility provides unattended (remote video) authentication, encryption, file decompression, and decryption. The Gemini system remotely validates visual monitoring and verifies that the video images are authentic.

The advantages of this system include reduced worker radiation exposure and reduced intrusion to facility operations. The first remote exchange of data and images occurred between U.S. and Russian weapons-usable nuclear material storage vaults. The Department of Energy has also been involved in the installation of remote monitoring systems and the initiation of field trials in Argentina, Australia, Japan, Sweden, and the European Commission Joint Research Center in Ispra, Italy.

NTSB INVESTIGATION

At Sky Harbor Airport in Phoenix, Arizona, over 700 cameras are connected to four rack rooms through a fiber optic backbone. Each rack room manages the video from recording to multiplexing and routing the images to any of eight locations. System operators utilize a custom graphical user interface (GUI) that displays maps of all areas, including camera locations. Operators can view live camera images, control pan/tilt/zoom, iris, and focus, view real-time recorded video from the previous eight hours, and view archived video from previous months.

The project was originally designed to monitor parking areas to help reduce car theft, assaults, and vandalism. The system was also used to monitor baggage claim areas. Activity in both areas was monitored and recorded in an effort to reduce the insurance liability for the City of Phoenix, which is self-insured. Since its installation the system has been expanded for use to view and record airfield activity, missing person searches, and theft and has even been instrumental in a National Transportation and Safety Board investigation.

The system was operating at peak efficiency when, during a routine landing, an America West A320 Airbus crashed on the north runway. Typically, the pilot knows if there's a problem with the landing gear, and he issues an alert so that the airport can prepare for the crash landing. In this situation, the pilot had no forewarning of equipment failure. As the incident unfolded, two of the 425 cameras installed on one of the airport parking garages were trained on the runway and recorded the crash from two separate angles. The video clearly shows the plane touching down, the failure in the front landing gear, and the subsequent skid to a stop at the edge of the runway. The NTSB is using the recorded video to help determine the cause of the crash.

WINEMAKERS WATCH THEIR VINES

The use of technology to assist winemakers is increasing, particularly in the United States and Australia, where wine grapes are often farmed in large, flat tracts of land where there is little change in weather and soil conditions. California winemakers can see how their vines are doing, day or night, with a wireless, web-based video system. Wineries buying grapes from Scheid Vineyards in Monterey, California can now access real-time pictures and data on their vines as they grow. The system allows about forty wineries across California to get a bird's eye view of the vines from three live, solar powered cameras located in the vineyards using a wireless network that covers 5,600 acres. The cameras can be operated remotely over the Internet and PTZ operated for close-up viewing. The network allows field managers to respond immediately to changing conditions and helps keep customers in touch with the growing process.

VIDEO MONITORING CONTRIBUTES TO GEOLOGICAL STUDIES

The USGS (United States Geological Survey) has used remote digital video for monitoring the variability in coastal sections. Characterizing the changes in shoreline positions and other physical changes has traditionally been a labor and cost intensive process, but remote video monitoring methods allow continuous sampling and can be maintained for extended periods of time.

The USGS has joined forces with Oregon State University to develop a program that provides inexpensive systems of data collection and analysis. Three field stations were set up to support projects in Southwest Washington, West Central Florida, and Lake Erie. The Florida station was designed to provide daily maps of shoreline evolution of a recently nourished region using software that was developed to convert video images to meaningful shoreline maps and historical descriptions.

KEEPING AN EYE ON THE COWS

A video system is being used by a large U.S. dairy farm to support the management of its dairy operations. The facility has 10,000 cows in four parlors and is located on 17,000 acres. It produces approximately 100,000 gallons of milk per day. Cameras view milking operations with two PTZ cameras, and three fixed cameras monitor the operation. A remote operator oversees whether cows are being prepped and handled properly. The facility also plans to use cameras at a visitor's center to be built in the next few years, where video will be displayed on a large screen so that visitors can see real time milking operations as they learn about it.

The cameras communicate video over a LAN to the main office located four miles away, where management can view operations at any time. They can also look in on milking operations from either home or offices at other locations. Other uses for cameras in dairy operations include the monitoring of the maternity area, feed storage, and mixing locations.

THE BUREAU OF ENGRAVING AND PRINTING

Can you think of a bigger challenge than providing security for the facilities that actually manufacture all of America's money? The Bureau of Engraving and Printing (BEP) was established in 1861 and in 1877 became the sole maker of all United States currency. Today, it is the largest producer of U.S. government security documents and prints billions of notes (bills) for delivery to the Federal Reserve System each year from production facilities in Washington, D.C. and in Fort Worth, Texas.

The Bureau designs, prints, and furnishes a variety of products, including Federal Reserve notes, U.S. postage stamps, Treasury securities, identification cards, naturalization certificates, and special security documents, and even does print runs for White House invitations and other such announcements. All documents with an associated dollar value are designed with advanced counterfeit deterrence features to ensure product integrity, and the process of printing is done under the scrutiny of a state of the art surveillance system. BEP procedures require extensive background checks for personnel hiring and strict security practices on the job. The digital video system has become an additional layer of confidence that procedures are followed and the facility remains highly secure.

The BEP has a security team monitoring cameras 24 hours a day, seven days a week, every day of the year. They also have remote viewing capabilities for authorized users, so they have the ability to look in any time, day or night. Security monitoring personal are not necessarily looking for people slipping twenty dollar bills into their pockets. In fact, there have been no successful attempts involving staff pilfering the merchandise since 1998. Instead, they are looking for abnormalities or inconsistencies in everyday operations of the production facilities, as well as violations in procedures. The system also does a great service to the employees of the BEP as errors are quickly resolved and an employee who might otherwise look suspicious is quickly cleared by video evidence.

For example, the job of a currency examiner is to retrieve packets of one hundred dollar bills from a conveyer belt and check them for printing errors. Personnel monitoring the currency examiner will make a report if a packet is missed or any other error occurs on the production line. Another interesting and closely monitored position is the person who checks individual currency notes, which are returned to the Bureau as unusable. After inspection and documentation by the examiner, these bills are destroyed. As one can imagine, destroying currency—even damaged bills—would be a tough job. The digital system ensures that proper procedures are followed and examiners are indeed destroying the damaged notes.

The system has also been used to investigate and solve Automatic Teller Machine (ATM) transaction disputes. The ATM located within the facility for the use of employees has a camera that records all transactions. The camera had been removed for repair and an obvious empty spot was in its place. Thinking the missing camera meant no video proof, a man denied having received cash from the machine. Unfortunately (for him), other cameras in the area recorded a perfect view of him taking his money from the machine during the time in question.

TRAFFIC MONITORING

The monitoring and control of traffic has taken many turns and with new technologies has become a sophisticated and progressive enterprise. In the past, electronic traffic management and information programs relied on a system of sensors for estimating traffic parameters. The more prevalent technology for this purpose was that of magnetic loop detectors buried underneath highways to count vehicles passing over them.

Various types of aircraft have also been used to monitor traffic, mostly in the form of television or radio station helicopters. These periodic flights over main arteries can communicate live traffic conditions but are usually limited to rush hour or specific traffic incidents and are of no real value in monitoring or controlling traffic on a 24 hour basis.

New digital video capabilities make it extremely viable for traffic monitoring, and it provides a number of advantages over older methods. A much larger set of traffic parameters can be monitored such as vehicle counts, types, and speeds as well as causes of congestion and recurrent accidents. In addition, traffic can now be monitored continuously.

16

New Roles of Digital Video

In 1997, the same agency responsible for the Internet, DARPA, began a three-year program to advance Video Surveillance and Monitoring (VSAM) technology. The purpose was to develop automated video understanding (intelligence) technology for use in future urban and battlefield surveillance applications. Advances resulting from this program allow a single human operator to monitor activities over a broad area using a distributed network of active video sensors.

The Carnegie Mellon University Robotics Institute and the Sarnoff Corporation teamed up to develop an end-to-end test bed system, which demonstrated a wide range of advanced surveillance techniques such as real-time moving object detection, tracking from stationary and moving camera platforms, recognition of object classes and specific object types among many other advanced analysis. Twelve other contracts were awarded for research in the areas of human activity recognition, vehicle tracking and counting, airborne surveillance, novel sensor design, and geometric methods for graphical view transfer. Today's digital video surveil-lance systems face the same difficulties that DARPA was assigned to overcome with the VSAM project: an overload of video informa-tion transmitted for view and response. A large surveillance system with 300 hundred cameras would need 13 operators to view every camera one time every 60 seconds with a sequencing system switching four monitors per operator every ten seconds. The primary function of a video analysis feature is to relieve CCTV operators from the stress of monitoring many screens of informa-tion that may not change for long periods. Even a moderately sized system containing eight cameras could prove impossible for an operator to monitor. Eight monitors could not be viewed with any degree of concentration for more than about 20 minutes. If the monitors were set to sequence, then activity on seven cameras is lost for most of the time and would be totally ineffective to detect intruders. This leaves too much time between images for adequate surveillance, and the fatigue factor inherent with this kind of stim-ulation is extreme.

The solution to this problem is either more staff to monitor the video, which could become cost prohibitive, or a system that is intelligent enough to detect a problem and signal for a response. Such a system would be considered to have artificial intelligence, which is defined as intelligence exhibited by an artificial entity.

Computer vision is a subfield of artificial intelligence with the purpose of programming a computer to understand the con-tents of an image. This ability is considered a class of artificial intelligence, which is basically a machine performing activities normally thought to require intelligence. Computer vision was actually developed in the 1950s and has been in use for some time, especially in the role of monitoring production lines. Now it is being used for security applications. More methods of video intel-ligence currently in use or being researched follow.

IMAGE ANALYSIS

Image analysis or video analysis involves the extraction of infor-mation from digital images by a method known as digital image processing. Image analysis can include from simple tasks like bar code reading to the much more complicated processes of facial recognition. Content analysis is simply a systematic analysis of the content rather than the structure of a communication. For example, the algorithm used concerns itself only with the shape of moving objects within a scene. These shapes are analyzed and classified. Actions are put into motion based on the resulting classification.

PATTERN RECOGNITION

Pattern recognition involving images occurs when raw data is reviewed by the computer and an action is put into motion based on the category of the data found. Most of us use a form of pattern recognition when we use an automatic spam filter. For example, a predefined pattern of data such as the word "drug" is acknowledged and an action is put into motion. In most cases the action is to block the spam. This is why so many e-mails are sent using false references like "returning your message" or "concerning your account."

PEOPLE COUNTING

Automatically counting people within a defined area and providing information about the direction of movement generates accurate traffic information that can enable efficient staffing, better queue control, and marketing data. This feature can also be set to alert staff when a safety threshold has been reached pertaining to occupancy limits.

INTELLIGENT VIDEO

Intelligent video refers to the analysis and extraction of video information with specific reasoning attached for specific applications. It may be determined that specific information is not of high enough priority, so resolution is decreased to conserve bandwidth or storage space. Intelligent video can also refer to content management, such as indexing and retrieval. It could also enhance an image by removing noise, improving resolution, or increasing the dynamic range.

Remember that we described algorithms as a kind of recipe or set of instructions? This explains why there are so many different capabilities offered by so many different companies. Most manufacturers focus on one or two specialties like identification, tracking, or determining motives for movements and incorporate those into their systems. Then they focus on niche markets that can best utilize those individual capabilities.

The intelligent video system can be programmed for an endless variety of uses. For example, a particular problem often found among train and bus stations is the gathering of loiterers. This problem is alleviated by system alarms that are instigated by the video itself when the presence of persons where they should not be is indicated. Personnel who have been alerted to the scene can view the area and determine the cause for the alarm, then dispatch someone to disperse the loiterers if necessary.

A similar feature is used to eliminate illegally parked cars from sensitive areas or abandoned vehicles. Automatic detection and alert for vehicles parked in restricted areas gives security personal immediate notice of a breach that could eliminate a potential car bomb threat or simply result in a parking ticket.

Some behavior recognition programs utilize recursive adaptive computer algorithms. These are dynamic programs through which normal patterns of behavior are continually learned and updated to define acceptable behavior patterns. These types of programs are useful because they can be programmed to alert irregularity in behavior patterns as they occur, not just those that have been previously identified.

FACIAL RECOGNITION

A facial recognition system is a computer driven application for automatically identifying a person from a digital image. It does that by comparing selected facial features in the live image and a facial database. A facial recognition system analyzes images of human faces through their special characteristics, such as the distance between the eyes, the length of the nose, and the angle of the jaw, etc. These characteristic features are called eigenfaces in the facial recognition domain.

The software compares images producing a score that measures similarities. Because the image is analyzed by individual characteristics the system can distinguish the same person with different appearances; for example, with or without glasses, shorter, longer, or even different color of hair, and seasonal skin color changes. None of these possible alterations change the dynamic structure of the face, meaning it remains identifiable.

Not all attempts at using facial recognition have achieved rave reviews; for example, the 2001 Super Bowl at Tampa Bay had very mixed reactions, but the technology continues to be refined and false positive responses are on the decline.

THE EVIDENTIARY DEBATE

The digitization of information and the resulting growth of computing and electronic networking bring a new kind of security threat. Not only must video information be kept from easily being destroyed, it must also be protected from any form of manipulation. Forensic video analysis has been accepted by the courts in reported cases since 1992. Organizations such as the International Association for Identification (IAI) have created resolutions incorporating digital images and data into their procedures. Formed in 1915, the IAI is the oldest and largest forensic science identification organization in the world.

A means of video authentication is necessary to use video as evidence in a court of law. Video authentication can be accomplished by creating a digital "fingerprint" for each digitized video clip. Techniques for safeguarding video integrity include technologies such as date/time stamping, watermarking, and encryption.

Encryption is simply the conversion of data into a form that is virtually impossible to understand by unauthorized people.

This provides a way to protect the privacy, security, confidentiality, integrity, and authenticity of wire and electronic communications, which can include video data. Decryption is the process of converting encrypted data back into its original form so it can be understood. In order to recover the contents of an encrypted signal, a decryption key is required. The key is an algorithm that reverses the work of the encryption algorithm. The longer the key, the harder it is to decipher the code.

Encryption software can use keys in different ways. For example, single-key encryption allows both the sender and receiver to use the same key to encrypt and decrypt messages. Depending on whether the encryption and decryption keys are the same or not, the process is called symmetric or asymmetric. A watermark can be embedded into an image to facilitate fingerprinting, authentication and integrity verification, content labeling, usage control, and content protection. Watermarking can be visible with the naked eye or may need special equipment to be seen.

Digital technology makes analysis and presentation of video evidence for the courts a reality, especially since court authority for electronic surveillance is not required by the Fourth Amendment for situations where there is no reasonable expectation of privacy. For example, no warrant is required to videotape activity in a parking lot, a bank, or other public places.

One of the more famous instances of video as evidence in the United States involved a video camera at an apartment complex a block from the Alfred P. Murrah Federal Building in Oklahoma City. A camera caught the image of a Ryder truck shortly before an explosion on April 19, 1995. It was later determined the explosion was caused by a homemade bomb hidden in the Ryder truck whose image was captured on camera.

This, more than any other incident before 9/11, has contributed to advancements and the increase in the use of CCTV as a surveillance and security tool in the United States.

Glossary

16 CIF: 16 × CIF (resolution 1408 × 1152).

4 CIF: 4 × CIF (resolution 704 × 576).

Analog: describes a continuous signal expressed as a continuous waveform.

Asynchronous: signal whose data is acknowledged or acted upon immediately, irrespective of any clock signal.

Aspect Ratio: ratio of width to height.

Bandwidth: measure of the carrying capacity of information over a network.

Baud: speed at which data is transmitted.

Binary: base-2 numbering system.

Bit: contraction of binary digit having two possible values; one or zero.

Bitmap: defines a display space and the color for each pixel or "bit" in the display space.

Byte: group of eight binary digits, or bits.

Cache: portion of RAM used for temporary storage of data needing rapid access.

CCD (Charge Coupled Device): semiconductor device (IC) that converts images to electronic signals.

CCIR (Comite Consulatif International Des Radiocommunications): European committee responsible for professional standards related to audio and video.

CCTV (Closed Circuit Television): television system used for private purposes, not for public or general broadcast.

Chroma Level: relating to the amount of saturation and hue at a particular point of an image.

Chrominance: color information contained in a video signal separate from the luminance component.

CIF (Common Intermediate Format): standard video formats defined by their resolution.

Coaxial Cable: standard cable consisting of a central inner conductor and a cylindrical outer conductor.

Codec: software that can compress a video source as well as play compressed video.

Color Bars: electronically generated video pattern consisting of eight equal width colors.

Color Burst: portion of a color video signal that contains a short sample of the color sub carrier used to add color to a signal.

Composite Video: electronic information needed to produce a video signal.

Conductor: material that allows electrical charges to flow through it.

Crosstalk: interference between two or more audio or video signals caused by unwanted stray signals.

Decimal: base-10 numbering system.

Digital Signal: signal that is either zero or one volt rather than a continuum of voltages or current.

DSP (Digital Signal Processor): primarily digital component used to process either digital or analog signals.

Distortion: degradation of a transmitted signal.

EMI (Electromagnetic Interference): signal impairment resulting from electromagnetic disturbances in the atmosphere.

Encryption: process applied to digital signals to ensure secure communications.

Ethernet: type of LAN that is recognized as an industry standard.

Field: single scan of a TV or monitor screen consisting of 262.5 lines. Two fields make up a single frame.

Frame: a complete video picture.

Giga: unit qualifier (symbol = G) representing one thousand million.

Gigabit: one billion bits.

Grayscale: An image type that uses black, white, and a range of shades of gray.

Hertz (Hz): unit of frequency; one Hertz equals one cycle, or one oscillation, per second.

HTML (Hypertext Markup Language): set of symbols or codes used in a file intended for display on a World Wide Web browser.

HTTP (Hypertext Transfer Protocol): set of rules for exchanging files (text, graphic images, sound, video, and other multimedia files) on the World Wide Web.

Internet: public network of computers and people sharing information. Anyone can access the Internet through an Internet service provider.

Intranet: private network of computers using web-based technology not accessible by the general public.

IP (Internet Protocol): method by which data is sent from one computer to another over the Internet.

IP Address: used to identify a particular computer on a network to other computers.

ISDN (Integrated Services Digital Network): international telecommunications standard.

ISO (International Organization for Standardization): nongovernment organization promoting the development of standardization to facilitate the international exchange of goods and services.

Kilo: unit qualifier (symbol = K) representing one thousand.

Kbs or Kbps: kilobits per second.

LAN (Local Area Network): multiple computers connected together to share information such as e-mail, files, and printers.

Luminance: lightness of a color measuring the intensity of light per unit area of its source, also called Luma.

Mbs or Mbps: megabits per second.

Mega: unit qualifier (symbol = M) representing one million.

Modem (Modulator/Demodulator): sends digital signals over analog lines.

Network: computers connected together for the purpose of sharing information.

Noise: disturbance in a signal.

NTSC (National Television Systems Committee): American television standard.

Peta: unit qualifier (symbol = P) representing one thousand million million.

Plasma: gaseous state in which the atoms or molecules are dissociated to form ions.

Point-to-multipoint: communications from one location to several locations.

Point-to-point: communications between two locations.

Quantization: reduce the number of colors or shades of gray in an image, with the goal being to reduce file size.

QCIF (Quarter CIF): resolution 176×144.

RAM (Random Access Memory): data-storage device from which data can be read out and new data can be written in.

ROM (Read Only Memory): "built-in" computer memory containing data that normally can only be read, not written on.

Router: device that connects two networks by reading the destination address of information sent over a network and forwarding it to the next step in its route.

Saturation: strength of a color with respect to its value.

Server: computer and software that provides some service for other computers connected to it through a network.

SQCIF (Sub Quarter CIF): resolution 128 × 96.

Tera: Unit qualifier (symbol = T) representing one million million.

TCP/IP (Transmission Control Protocol/Internet Protocol): basic communication language of the Internet.

Twisted Pair: electrical conductor that consists of two wires twisted around each other.

Value: brightness of a color, based on the amount of light emanating from it.

VDU (Visual Display Unit): cathode ray unit designed for display of video pictures.

Video: signal output from a video camera.

WAN (Wide Area Network): multiple LANs connected over a distance that share information.

Wavelet: mathematical function useful in signal processing and image compression.

Wireless: electromagnetic waves, such as radio or television, that carry a signal from one section to another.

Index

BONUS GLOSSARY

SURVEILLANCE TERMS

Excerpted from
**Herman Kruegle, *CCTV Surveillance: Analog and Digital Video
Practices and Technology, Second Edition***
ISBN: 978-0-7506-7768-4
December 2006

Have you seen the new, fully updated edition of this classic Butterworth-Heinemann Security title?

This revision offers extensive coverage of one of the most common applications of digital CCTV. Herman Kruegle bring years of experience with design, installation, and management of CCTV systems to bear on the new issues created by increasingly complex technologies.

Now that you've read Harwood's *Digital CCTV*,
look for *CCTV Surveillance, Second Edition*, and take your
knowledge to the next level!

GO TO
http://books.elsevier.com/security
FOR COMPLETE INFORMATION ON
ORDERING BUTTERWORTH-HEINEMANN
SECURITY BOOKS

Glossary of CCTV Surveillance Terms

Many terms and definitions used in the security industry are unique to CCTV surveillance; others derive from the electro-optical and information-computer industries. This comprehensive glossary will help the reader better understand the literature, interpret manufacturers' specifications, and write bid specifications and requests for quotation. These terms encompass the CCTV, physical computer and communications industries, basic physics, electricity, mechanics, and optics.

Aberration Failure of an optical lens to produce exact point-to-point correspondence between an object and its image.

ABC Automatic brightness control In display devices, the self-acting mechanism that controls brightness as a function of ambient light.

Access point An electronic device for connecting wireless PC cameras directly to the Internet.

Achromatic lens A lens consisting of two or more elements, usually of crown and flint glass, that has been corrected for chromatic aberration with respect to two selected colors or light wavelengths.

Active Video Lines   The video lines producing the picture. All video lines not occurring in the horizontal and vertical blanking intervals.

ADSL Asymmetric DSL   A DSL technology providing asymmetrical bandwidth over a single wire pair. The downstream bandwidth going from the network to the subscriber is typically greater than the upstream bandwidth going from the subscriber to the network. See **Direct Subscriber Line**.

AF Auto-focus   A system by which the camera lens automatically focuses on a selected part of the video scene.

AFC Automatic frequency control   A feature whereby the frequency of an oscillator is automatically maintained within specified limits.

AGC Automatic gain control   A process by which gain is automatically adjusted as a function of input or other specified parameter to maintain the output nearly constant.

Alarming switcher   see **Switcher, Alarming**.

Ambient temperature   The temperature of the environment. The temperature of the surrounding medium, such as air, gas, or liquid that comes into contact with the apparatus.

Amplifier   A device whose output is essentially an enlarged reproduction of the input and that does not draw power from the input source.

Amplifier, distribution   A device that provides several isolated outputs from one looping or bridging input. The amplifier has sufficiently high input impedance and input-to-output isolation to prevent loading of the input source.

Analog Signal   The representation of data by continuously variable quantities in digital format as opposed to a finite number of discrete quantities. An electrical signal that varies continuously, not having discrete values.

Analog Television   The "standard" television broadcast. Analog signals vary continuously, representing fluctuations in color and brightness of a visual scene.

Angle of view   The maximum scene angle that can be seen through a lens or optical-sensor assembly. Usually described in degrees, for horizontal, vertical, or circular dimension.

Antenna   An electrical signal gathering or transmitting device used with electrical receivers and transmitters for collecting or propagating an electrical signal through the airwaves. The primary antenna specifications are vertical and horizontal directivity and gain, input impedance and bandwidth, and power handling capacity.

Aperture   An opening that will pass light, electrons, or other forms of radiation. In an electron gun, the aperture determines the size of, and has an effect on, the shape of the electron beam. In television optics, the aperture is the effective diameter of the lens that controls the amount of light reaching the image sensor.

Aperture, clear   see **Clear aperture**.

Aperture, numerical   see **Numerical aperture**.

Aperture stop   An optical opening or hole that defines or limits the amount of light passing through a lens system. The aperture stop takes the form of the front lens diameter in a pinhole lens, an iris diaphragm, a neutral density or spot filter.

Arc lamp   An electric-discharge lamp with an electric arc between two electrodes to produce illumination. The illumination results from the incandescence of the positive electrode and from the heated, luminous, ionized gases that surround the arc.

ASIS   American Society for Industrial Security.

Aspect ratio   The ratio of width to height for the frame of the video picture in CCTV or broadcast television. The NTSC and PAL standard is 4:3. The aspect ratio for high-definition television HDTV is 16:9.

Aspheric   An optical element having one or more surfaces that are not spherical. The spherical surface of the lens is slightly altered to reduce spherical aberration, thereby improving image quality.

Astigmatism   A lens aberration that causes an object point to be imaged as a pair of short lines at right angles to each other.

ATM Asynchronous Transfer Mode   The communications standard that allows multiple traffic types (voice, video, or data) to be conveyed in fixed-length cells rather than the random length "packets" as used in Ethernet and Fiber Distributed Data Interface (FDDI). This enables very high speeds, making ATMs popular for demanding network backbones. Newer ATMs also support WAN transmissions.

ATSC Advanced Television Systems Committee   An international organization with a committee responsible for digital television standards and development.

Attenuation   A reduction in light or electrical signal or energy strength. In electrical systems attenuation is often measured in decibels or decibel per unit distance. In optical systems the units of measure are f-number or optical density. See also **Decibel**.

Audio frequency   Any frequency corresponding to a normally audible sound wave—roughly from 15 to 15,000 Hz.

Auto balance   A system for detecting errors in color balance in the white and black areas of the picture and automatically adjusting the white and black levels of both the red and blue signals as needed.

Auto light range   The range of light—such as sunlight to moonlight or starlight—over which a TV camera is capable of automatically operating at specified output and within its dynamic range.

Automatic iris   A mechanical diaphragm device in the lens that self-adjusts optically to light level changes via the video signal from the television camera. The iris diaphragm opens or closes the aperture to control the light transmitted through the lens. Typical compensation ranges are 10,000–300,000 to 1. Automatic iris lenses are used on solid-state CCD, ICCD, and CMOS cameras and SIT, ISIT tube cameras.

Automatic iris control   An electro-optic accessory to a lens that measures the video level of the camera and opens and closes the iris diaphragm to compensate for light changes.

Automatic light compensation   The degree to which a CCTV camera can adapt to varying lighting conditions.

Automatic light control   The process by which the illumination incident upon the face of a pickup device is automatically adjusted as a function of scene brightness.

Automatic sensitivity control   The self-acting mechanism that varies system sensitivity as a function of specified control parameters. This may include automatic target control, automatic light control, etc. or any combination thereof.

Axis, optical   see **Optical axis**.

Backbone   The part of a network that acts as the primary network path for traffic moving between rather than within networks.

Back focus   The distance from the last glass surface of a lens to the focused image on the sensor.

Back porch   That part of a composite video signal that lies between the trailing edge of a horizontal sync pulse and the trailing edge of the corresponding blanking pulse. The color burst, if present, is not considered part of the back porch.

Band pass   A specific range of frequencies that will be passed through an optical or electronic device or system.

Bandwidth   The data carrying capacity of a device or network connection. The number of hertz (Hz, cycles per second) expressing the difference between the lower and upper limiting frequencies of a frequency band. Also, the width of a band of frequencies.

Bandwidth limited gain control   A control that adjusts the gain of an amplifier while varying the bandwidth. An increase in gain reduces the bandwidth.

Barrel distortion   An electronic or optical distortion in television systems that makes the video image appear to bulge outward on all sides like a barrel.

Beam   A concentrated, unidirectional flow of electrons, photons, or other energy: 1) A shaft or column of light; a bundle of rays consisting of parallel, converging, or diverging rays. 2) A concentrated stream of particles that is unidirectional. 3) A unidirectional concentrated flow of electromagnetic waves.

Beam splitter   An optical device for dividing a light beam into two or more separate beams. The splitting can be done in the parallel (collimated) beam or in the focused image plane.

Beam width, angular beam width   The angular beam width of a conical beam of light. The vertex angle of the cone, which determines the rate at which a beam of energy diverges or converges. Lasers produce very narrow angle or very nearly parallel beams. Thermal light sources (filament, fluorescent, etc.) produce wide-angle beams.

Beta format   An original 1/2-inch Sony video cassette recorder format not compatible with the VHS format.

Bifocal lens A lens system having two different focal length lenses that image two identical or different scenes onto a single camera sensor. The two scenes appear as a split image on the monitor.

Bit binary digit The smallest unit of computer memory. A bit is a binary digit representing two different states (1, 0), either on or off. A method of storing information that maps an image bit by bit.

Bitmap A bitmap defines a display space and the color for each pixel or "bit" in the display space. GIF and JPEG are examples of graphic image file types that contain bitmaps.

Bit rate The bit rate is the number of bits transmitted per second (bps).

Blackbody A thermally heated body that radiates energy at all wavelengths according to specific physical laws.

Black clamp An electronic circuit that automatically maintains the black video level (no light) at a constant voltage.

Black level The picture signal level corresponding to a specified maximum limit for black peaks.

Black negative The television picture signal in which the polarity of the voltage corresponding to black is negative with respect to that which corresponds to the white area of the picture signal.

Blanking The process whereby the beam in an image pickup or cathode ray display tube is cut off during the retrace period so that it won't create any visible information on the screen.

Blanking level The level of a composite picture signal that separates the range containing picture information from the range containing synchronizing information; also called pedestal, or blacker-than-black.

Blooming   In a camera, the visual effects caused by over exposing a CCD or other sensor. In CRT monitors, the condition that occurs when the phosphors on the screen are driven harder than they should be. This causes defocusing of regions of the picture where the brightness is at an excessive level, due to enlargement of spot size and halation of the screen of the CRT tube.

BNC Bayonet-Neil-Concelman   A connector named after its designers and widely used in the video and RF transmission industry for terminating and coupling coaxial cables. The BNC connector is easy to install, reliable, and with little video signal loss. Used with 75 ohm cable for video applications cable and 50 ohm cable for RF.

Borescope   An optical device used for the internal inspection of mechanical and other parts. The long tube contains a multiple lens telescope system that usually has a high f-number (low amount of light transmitted).

Boresight   An optical instrument used to check alignment or pointing direction. A small telescope mounted on a weapon or video camera so that the optical axis of the telescope and the mechanical axis of the device coincide. The term also applies to the process of aligning other optical equipment.

Bounce   Sudden variation in picture presentation (brightness, size, and so on) independent of scene illumination.

Breezeway   In NTSC color, that portion of the back porch between the trailing edge of the sync pulse and the start of the color burst.

Bridge   A digital electronic device that passes data packets between multiple network segments using the same communication protocol. If a packet is destined for use within the sender's own network segment, the bridge keeps the packet local. If the packet is bound for another segment, the bridge passes the packet onto the network backbone.

Bridging Connecting two electrical circuits in parallel. Usually the input impedances are large enough so as not to affect the signal level.

Bridging amplifier An amplifier for bridging an electrical circuit without introducing an apparent change in the performance of that circuit.

Brightness The attribute of visual perception whereby an area appears to emit more or less light. Luminance is the recommended name for this photometric quantity, which has also been called brightness.

Brightness control The manual bias control on a cathode ray tube or other display device that determines both the average brightness and the contrast of a picture.

Browser An application program that provides a way to look at and interact with all the information on the World Wide Web. The two most popular Web browsers are Netscape and Microsoft Internet Explorer.

Buffer A temporary computer storage area usually held in RAM and used as a temporary holding area for data.

Burn-in Also called burn. An image that persists in a fixed position in the output signal of a camera tube after the camera has been pointed toward a different scene. An image that persists on the face of a CRT monitor with no input video signal present.

Byte A group of 8 bits or 256 discrete items of information, such as color, brightness, etc. The basic unit of information for the computer.

C-mount An industry standard for lens mounting. The C-mount has a thread with a 1-inch diameter and 32 threads per inch. The distance from the lens mounting surface to the sensor surface is 0.69 inches (17.526 mm).

Cable A number of electrical conductors (wires) bound in a common sheath. These may be video, data, control, or voice cables. They may also take the form of coaxial or fiber-optic cables.

Cable Modem A class of modem that is used for connecting to a cable TV network, which in turn can connect directly to the Internet. Cable based connections to the Internet are typically much faster than dial-up modems.

Camera control unit CCU Remote module that provides control of camera electronic circuitry such as camera shutter speed, video amplification, and lens parameters.

Camera format Video cameras have 1/6, 1/4, 1/3, 1/2, and 2/3-inch sensor image formats. The actual scanned areas used on the sensors are 3.2 mm horizontal × 2.4 mm vertical for the 1/4-inch, 4.8 mm horizontal × 3.6 mm vertical for the 1/3-inch, 6.4 mm horizontal × 4.8 mm vertical for the 1/2-inch, and 8.8 mm horizontal × 6.6 mm vertical for the 2/3-inch.

Camera housing An enclosure designed to protect the video camera from tampering or theft when indoors or outdoors or from undue environmental exposure when placed outdoors.

Camera, television An electronic device containing a solid state sensor or an electronic image tube and processing electronics. The image formed by a lens ahead of the sensor is clocked out for a solid state sensor or rapidly scanned by a moving electron beam in a tube camera. The sensor signal output varies with the local brightness of the image on the sensor. These variations are transmitted to a CRT, LCD, or other display device, where the brightness of the scanning spot is controlled. The scanned location (pixel) at the camera and the scanned spot at the display are accurately synchronized.

Camera tube An electron tube that converts an optical image into an electrical current by a scanning process. Also called a pickup tube or television camera tube.

Candela cd Unit of measurement of luminous intensity. The candela is the international unit that replaces the candle.

Candle power, cp Light intensity expressed in candles. One foot-candle (fc) is the amount of light emitted by a standard candle at 1-foot distance.

Catadioptric system A telephoto optical system embodying both lenses and image-forming mirrors. Examples are the Cassegrain, Schmidt and Maksutov telescope. Mirrors are used to reduce the size and weight of these long focal length lenses.

CAT Cable A class of cables using unshielded twisted pairs (UTP) for transmitting video, audio, data, and controls. The differences are based mainly on bandwidth, copper size, and electrical performance. The most common are CAT 3, 4, 5, 5e, and 6, as defined by the EIA and TIA (Telecommunications Industry Association).

Cathode ray tube CRT A vacuum tube in which electrons emitted by a heated cathode are focused into a beam and directed toward a phosphor-coated surface, which then becomes luminescent at the point where the electron beam strikes. Prior to striking the phosphor, the focused electron beam is deflected by electromagnets or two pairs of electrostatically charged plates located between the electron "gun" and the screen.

CATV Cable television, Community antenna television A cable television distribution system primarily used for consumer TV broadcast programming to a building or small community.

CCD Charge coupled device A solid-state semiconductor imaging device used in most current security cameras. This sensor has hundreds of thousands of photo-sites (pixels) that convert light energy into electronic signals and with the camera electronics eventually into the video signal. The CCD sensor sizes is used in security systems are: 1/6, 1/4, 1/3, and 1/2 inch (measured diagonally).

CCIR International Radio Consultative Committee A global organization responsible for establishing television standards. The CCIR format uses 625 lines per picture frame, with a 50 Hz power line frequency.

CCTMA Closed Circuit Television Manufacturers Association A former division of the EIA, a full-service national trade organization promoting the CCTV industry and the interests of its members.

CCTV camera The part of a CCTV system that captures a scene image and converts the light image into an electrical representation for transmission to a display or recording device.

CCTV Closed-circuit television A closed television system used within a building or complex to visually monitor a location or activity for security or industrial purposes. CCTV does not broadcast consumer TV signals but transmits in analog or digital form over a closed circuit via an electrically conducting cable, fiber-optic cable, or wireless transmission.

CCTV monitor That part of the CCTV system which receives the picture from the CCTV camera and displays it.

Character generator The equipment used to create titles or other text in a video image. The device is used to generate words and numbers in a video format.

Chroma The characteristics of color information, independent of luminance intensity. Hue and saturation are qualities of chroma. Black, gray, and white objects do not have chroma characteristics.

Chromatic aberration A design flaw in a lens or lens system that causes the lens to have different focal lengths for radiation of different wavelengths. The dispersive power of a simple positive lens focuses light from the blue end of the spectrum at a shorter distance than light from the red end. This deficiency produces an image that is not sharp.

Chrominance signal   The portion of a video signal that carries hue and saturation color information.

CIF Common Image Format   A standard defining a digital pixel resolution. In the HDTV format the CIF resolution is 1,920 × 1,080 pixels, not to be confused with Common Intermediate Format.

CIF Common Intermediate Format   A commonly used television standard for measuring resolution. One CIF equals 352 × 240 pixels for NTSC and 352 × 288 for PAL. Full resolution is considered to be 4CIF, which is 704 × 480 pixels for NTSC and 704 × 576 pixels for PAL. Quarter resolution or 1/4CIF equals 176 × 120 for NTSC and 176 × 144 for PAL.

Clamping   The process and circuitry that establishes a fixed level for the television picture level at the beginning of each scanning line.

Clear aperture   The physical opening in a lens or optical system that restricts the extent of the bundle of rays incident on the given surface. It is usually circular and specified by its diameter.

Client   A networked PC or terminal that shares "services" with other PCs. These services are stored on or administered by a server.

Client/Server   Client/Server describes the relationship between two computer programs in which one program, the client, makes a service request from another program, the server, which fulfills the request. The client/server model is one of the founding concepts of network computing. Most business applications written today use the client/server model as does the Internet's main program, TCP/IP.

Clipping   The shearing off of the peaks of a signal. For a picture signal, clipping may affect either the positive (white) or negative (black) peaks. For a composite video signal, the sync signal may be affected.

Close-up lens   A low-magnification (low power) accessory lens that permits focusing on objects closer to the lens than it has been designed for.

CMYK   The primary *colors* of light are red, green, and blue. The primary colors of *pigments* such as ink or paint are cyan, magenta, and yellow. Adding all three pigments together should produce black but often produces a poor black. Therefore black is obtained from sources such as carbon and called the fourth "primary" in the printing process. The letter K is used for black. The grouping of the letters CMYK is usually associated with the color print industry.

Coaxial cable   A cable capable of carrying a wide range of frequencies with low signal loss. In its simplest form it consists of a stranded metallic shield with a single wire accurately placed along the center of the shield and isolated from the shield by an insulator.

Codec EnCOder/DECcoder   A process or device by which or in which a signal is encoded for transmission or storage, then decoded for play back. An algorithm that handles the compression and decompression of video files. As a device, a box or computer card that accomplishes the encode/decode process.

Collimated   A parallel beam of light or electrons.

Color bar test pattern   A special test pattern for adjusting color TV receivers or color encoders. The upper portion consists of vertical bars of saturated colors and white. The lower horizontal bars have black-and-white areas and I and Q signals.

Color saturation   The degree of mixture of a color and white. When a color is mixed with little or no white, it is said to have a high saturation. Low saturation denotes the addition of a great amount of white, as in pastel colors.

Color temperature   The term used to denote the temperature of a blackbody light source that produces the same color as the light under consideration. Stated in degrees Kelvin.

Component video   The uncoded output of a video camera, recorder, etc., whereby the red, green, blue, chrominance, and luminance signals are kept separate. Not to be confused with composite video.

COM port   A serial communication port that supports the RS-232 standard of communication.

Composite video   An analog, encoded signal used in television transmission, including the picture signal (intensity and color), a blanking signal, and vertical and horizontal synchronizing signals. All components of the composite video signal are transmitted down a single cable simultaneously.

Compression, Analog   The reduction in gain at one level of a analog signal with respect to the gain at another level of the same signal.

Compression, Digital   The removal of redundant information from a signal to decrease the digital transmission or storage requirements. The use of mathematical algorithms to remove redundant data (bits) at the sending end without changing its essential content (encoding). The two generic types are *lossy* and *lossless*. There are many compression algorithms with the most common being MPEG, M-JPEG, H.264, and wavelet.

Concave   A term describing a hollow curved surface of a lens or mirror; curved inward.

Contrast   The range of difference between light and dark values in a picture, usually expressed as contrast ratio (the ratio between the maximum and minimum brightness values).

Control panel   A rack at the monitor location containing a number of controls governing camera selection, pan and tilt controls, focus and lens controls, etc.

Convergence   The crossover of the three electron beams of a three-gun tri-color picture tube. This normally occurs at the plane of the aperture mask.

Convex   A term denoting a spherically shaped optical surface of a lens or mirror; curved outward.

Corner reflector, corner cube prism   A corner reflector having three mutually perpendicular surfaces and a hypotenuse face. Light entering through the hypotenuse is totally internally reflected by each of the three surfaces in turn, and emerges through the hypotenuse face parallel to the entering beam and returns entering beams to the source. It may be constructed from a prism or three mutually perpendicular front surface mirrors.

Covert surveillance   In television security, the use of camouflaged (hidden) lenses and cameras for the purpose of viewing a scene without being seen.

Cross-talk   Interference between adjacent video, audio, or optical channels.

CS-mount   An industry standard for lens mounting. The CS-mount has a thread with a 1-inch diameter and 32 threads per inch. The distance from the lens mounting surface to the sensor surface is 0.492 inches (12.497 mm).

Cutoff frequency   That frequency beyond which no appreciable energy is transmitted. It may refer to either an upper or lower limit of a frequency band.

Dark current   The charge accumulated by pixels while not exposed to light. The current that flows in a photo-conductor when it is placed in total darkness.

Dark current compensation   A circuit that compensates for the dark current level change with temperature.

DC restoration   The re-establishment by a sampling process of the DC and low-frequency components of a video signal that has been suppressed by AC transmission.

DC transmission   A form of transmission in which the DC component of the video signal is transmitted.

Decibel dB   A measure of the voltage or power ratio of two signals. In system use, a measure of the voltage or power ratio of two signals provided they are measured across the same value of impedance. Decibel gain or loss is 20 times log base 10 of the voltage or current ratio (Voutput/Vinput), and 10 times log base 10 of the power ratio (Poutput/Pinput).

Decoder   The circuitry in a receiver that transforms the detected signal into a form suitable to extract the original modulation or intelligence.

De-Compression   The process of taking a compressed video signal and returning it to its original (or near original) form it had before compression.

Definition   The fidelity of a television system with respect to the original scene.

Delay distortion   Distortion resulting from the nonuniform speed of transmission of the various frequency components of a signal, caused when various frequency components of the signal have different times of travel (delay) between the input and the output of a circuit.

Delay line   A continuous or periodic structure designed to delay the arrival of an electrical or acoustical signal by a predetermined amount.

Density   A measure of the light-transmitting or reflecting properties of an optical material. It is expressed by the common logarithm of the ratio of incident to transmitted light flux. A material having a density of 1 transmits 10% of the light, of 2 transmits 1%, of 3 transmits 0.1%, etc. See **Neutral density filter**.

Depth of field   The area between the nearest and the farthest objects in focus. For a lens, the area along the line of sight in which objects are in reasonable focus. It is the measured from the distance behind an object to the distance in front of the object when the viewing lens shows the object to be in focus. Depth of field increases with smaller lens aperture (higher f-numbers), shorter focal lengths, and greater distances from the lens.

Depth of focus   The range of detector-to-lens distance for which the image formed by the lens is clearly focused.

Detail contrast   The ratio of the amplitude of video signal representing high-frequency components with the amplitude representing the reference low-frequency component, usually expressed as a percentage at a particular line number.

Detail enhancement   Also called image enhancement. A system in which each element of a picture is analyzed in relation to adjacent horizontal and vertical elements. When differences are detected, a detail signal is generated and added to the luminance signal to enhance it.

Detection, image   In video, the criterion used to determine whether an object or person is observed (detected) in the scene. Detection requires the activation of only 1 TV line pair.

DHCP Dynamic Host Configuration Protocol   A protocol that lets network administrators automate and centrally manage the assignment of dynamic IP addresses to devices or a network. These are temporary addresses that are created anew for each transmission. The DHCP keeps track of both dynamic and static IP addresses, saving the network administrator the additional task of manually assigning them each time a new device is added to the network.

Diaphragm   see **Iris diaphragm**.

Differential gain   The amplitude change, usually of the 3.58-MHz color subcarrier, introduced by the overall video circuit, measured in dB or percent, as the picture signal on which it rides is varied from blanking to white level.

Differential phase   The phase change of the 3.58-MHz color sub carrier introduced by the overall circuit, measured in degrees, as the picture signal on which it rides is varied from blanking to white level.

Digital 8   A Sony format that uses Hi8 or 8 mm tapes to store digital video.

Digital signal   A video signal that is comprised of bits of binary data, otherwise known as ones and zeros (1, 0). The video signal travels from the point of its inception to the place where it is stored, and then on to the place where it is displayed either as an analog or digital presentation.

Digital Zoom   The process by which a camera takes a small geo-metrical part of the original captured frame and zooms it digitally with interpolation. Generally causes image degradation above 2× to 3× zoom ratios.

Diopter   A term describing the optical power of long focal length lenses. It is the reciprocal of the focal length in meters. For example, a lens with a focal length of 25 cm (0.25 m) has a power of 4 diopters.

Dipole antenna   The most common antenna used in video for wireless transmission of the analog or digital video signal consists of a 50 ohm coaxial with a length of exposed center conductor equal to a quarter wavelength of the transmission frequency. See also **Yaggi antenna**.

Direct-Sequence Spread Spectrum DSSS   see **Spread Spectrum**.

Distortion, electrical   An undesired change in the waveform from that of the original signal.

Distortion, optical   A general term referring to the situation in which an image is not a true reproduction of an object. There are many types of distortion.

Distribution amplifier   see **Amplifier, distribution**.

Dot bar generator   A device that generates a specified output pattern of dots and bars. It is used to measure scan linearity and geometric distortion in video cameras and monitors. Also used for converging cathode ray tubes.

DRAM Dynamic random access memory   A type of computer memory that is lost when the power is turned off.

Drive pulses   Synchronizing and blanking pulses.

DSL Direct Subscriber Line   A digital phone service that provides full voice, video, and digital data over existing phone systems at higher speeds than are available in typical dial-up Internet sessions. There are four types: ADSL, HDSL, SDSL, and VDSL. All operate via modem pairs: one modem located at a central office and the other at the customer site. Asymmetric DSL technology provides asymmetrical bandwidth over a single wire pair. The downstream bandwidth from network to the subscriber is typically greater than the upstream bandwidth from the subscriber to the network.

DSL Modem   A modem that connects a PC to a network, which in turn connects to the Internet.

DTV Digital Television   Refers to all formats of digital video, including SDTV and HDTV.

DVD   Originally called Digital video disks, now called digital versatile disk. These high-capacity optical disks now store everything from massive computer applications to full-length movies. Although similar in physical size and appearance to a CD or a CD-ROM the DVD significantly improves on its predecessors' 650 MB of storage. A standard single layer single-sided DVD can store 4.7 GB of data. The two layer, single-sided version boosts the capacity to 8.5 GB. The double-sided version stores 17 GB but requires a different disk drive for the PC.

DVR Digital video recorder   Records video pictures digitally.

Dynamic range   In television, the useful camera operating light range from highlight to shadow, in which detail can be observed in a static scene when both highlights and shadows are present. In electronics, the voltage or power difference between the maximum allowable signal level and the minimum acceptable signal level.

Echo   A signal that has been reflected at one or more points during transmission with sufficient magnitude and time difference as to be detected as a signal distinct from that of the primary signal. Echoes can be either leading or lagging the primary signal and appear on displays as reflections or "ghosts".

EDTV Extended definition television   A marketing term for a standard definition television set that displays a progressive scan (non-interlaced) picture and usually has a horizontal resolution near the high end of SDTV (over 600 pixels).

EIA Electronics Industry Association   EIA is a trade alliance for its members engaged in or associated with the manufacture, sale or distribution of many categories of electronic equipment.

EIA interface   A standardized set of signal characteristics (time duration, waveform, delay, voltage, current) specified by the Electronic Industries Association.

EIA sync signal   The signal used for the synchronizing of scanning specified in the Electrical Industry Association standards RS-170 (for monochrome), RS-170A (for color), RS-312, RS-330, RS-420, or subsequent specifications.

Electromagnetic focusing   A method of focusing a cathode ray beam to a fine spot by application of electromagnetic fields to one or more deflection coils of an electron lens system.

Electronic viewfinder   see **Viewfinder, electronic.**

Electron beam   A stream of electrons emitted from the cathode of a cathode ray tube.

Electrostatic focusing   A method of focusing a cathode ray beam to a fine spot by application of electrostatic potentials to one or more elements of an electron lens system.

Endoscope   An optical instrument resembling a long, thin periscope used to examine the inside of objects by inserting one end of the instrument into an opening in the object. Endoscopes comprise a coherent fiber-optic bundle with a small objective lens to form an image of the object onto one end of the bundle, and a relay magnifier lens at the sensor end to focus the fiber bundle image onto the sensor. In an illuminated version, light from an external lamp is piped down to the object by a second set of thicker fibers surrounding the image-forming bundle. See also **Fiberscope**.

Equalizer   An electronic circuit that introduces compensation for frequency discrimination effects of elements within the television system.

Ethernet   The most widely installed local area network technology that uses collision detection to move data packets between workstations. Originally developed by Zerox and then developed further by Digital Equipment Corp. (DEC) and Intel. An Ethernet local area network (LAN) typically uses coaxial or fiber optic cable, or twisted pair wires. The most common Ethernet is called 10BASE-T and provides transmission speeds up to 10 Mbps. See **Fast Ethernet, Gigabit Ethernet**.

Extranet   A network that provides external users (suppliers, independent sales agents, dealers) access to company documents such as price lists, inventory reports, shipping schedules, and more.

Fader   A control and associated circuitry for affecting fade-in and fade-out of video or audio signals.

Fast Ethernet   The 100BASE-T10 Ethernet that provides transmission speeds up to 100 Mbps, 10 times faster than 10BASE-T.

FCC Federal Communications Commission   The FCC is an independent federal regulatory agency charged with establishing policies to cover interstate and international communications via radio, television, wire, satellite, and cable.

FDDI Fiber Distributed Data Interface   A LAN technology based on a 100 Mbps token passing networking over fiber-optic cable. FDDI is usually reserved for network backbones in larger organizations.

Fiber-optic bundle, coherent   An optical component consisting of many thousands of hair-like fibers coherently assembled so that an image is transferred from one end of the bundle to the other. The length of each fiber is much greater than its diameter. The fiber bundle transmits a picture from one end of the bundle to the other, around curves and into otherwise inaccessible places by a process of total internal reflection. The positions of all fibers at both ends are located in an exact one-to-one relationship with each other.

Fiber-optic transmission   The process whereby light is transmitted through a long, transparent, flexible fiber, such as glass or plastic, by a series of internal reflections. For video, audio, or data transmission over long distances (thousands of feet, many miles) the light is modulated and transmitted over a single fiber in a protective insulating jacket. For light *image* transmission closely packed bundles of fibers can transmit an entire coherent image where each single fiber transmits but one component of the whole image.

Fiberscope   A bundle of systematically arranged fibers that transmits a monochrome or full-color image which remains undisturbed when the bundle is bent. By mounting an objective lens on one end of the bundle and a relay or magnifying lens on the other, the system images remote objects onto a sensor. See **Endoscope**.

Field   One of the two equal parts into which a television frame is divided in an interlaced system of scanning. There are 60 fields per second in the NTSC system and 50 in the CCIR and PAL systems. The NTSC field contains 262 1/2 horizontal TV lines and the CCIR, PAL 312.5 TV lines.

Field frequency   The number of fields transmitted per second in a television system. Also called field repetition rate. The U.S. standard is 60 fields per second (60-Hz power source). The European standard is 50 fields per second (50-Hz power source).

Field lens   A lens used to affect the transfer of the image formed by an optical system to a following lens system with minimum vignetting (loss of light).

Filter   An optically transparent material characterized by selective absorption of light with respect to wavelength (color). Electrical network of components to limit the transmission of frequencies to a special range (bandwidth).

FireWire   Also known as IEEE 1394 or i.LINK, FireWire is a two-way digital connection between computers and peripherals like digital camcorders and cameras. Most equipment uses 4-pin ports and connectors, but some use the 6-pin version.

FFL Fixed focal length lens   A lens having one or more elements producing a singular focal length. The focal length is measured in millimeters or inches.

FHSS Frequency Hopping Spread Spectrum   see **Spread Spectrum Modulation**.

Firewall   A set of programs that protects the resources of a private network from outside users.

Flatness of field   Appearance of the image to be flat and focused. The object is imaged as a plane.

Fluorescent lamp A high-efficiency, low-wattage arc lamp used in general lighting. A tube containing mercury vapor and lined with a phosphor. When current is passed through the vapor, the strong ultraviolet emission excites the phosphor, which emits visible light. The ultraviolet energy cannot emerge from the lamp as it is absorbed by the glass.

FM Frequency modulation A process of translating baseband information to a higher frequency. The process: the two input signals that are inputs to an FM modulator are the baseband signal (video) and the carrier frequency (a constant amplitude and constant frequency signal). The frequency of the carrier is modulated (i.e. changed) and increased and decreased about its center frequency by the amplitude of the baseband signal.

f-number The optical speed or ability of a lens to pass light. The f-number (f/#) denotes the ratio of the equivalent focal length (FL) of an objective lens to the diameter (D) of its entrance pupil (f/# = FL/D). The f-number is directly proportional to the focal length and inversely proportional to the lens diameter. A smaller f-number indicates a faster lens.

Focal length, FL The distance from the lens center, or second principal plane to a location (plane) in space where the image of a distant scene or object is focused. FL is expressed in millimeters or inches.

Focal length, back The distance from the rear vertex of the lens to the lens focal plane.

Focal plane A plane (through the focal point) at right angles to the principal axis of a lens or mirror. That surface on which the best image is formed.

Focal point The point at which a lens or mirror will focus parallel incident radiation from a distant point source of light.

Focus   (1) The focal point. (2) The adjustment of the eyepiece or objective of a visual optical device so that the image is clearly seen by the observer. (3) The adjustment of a camera lens, image sensor, plate, or film holder so that the image is sharp. (4) The point at which light rays or an electron beam form a minimum-size spot. Also the action of bringing light or electron beams to a fine spot.

Focus control, electronic   A manual electric adjustment for bringing the electron beam of an image sensor tube or picture tube to a minimum size spot, producing the sharpest image.

Focus control, mechanical   A manual mechanical adjustment for moving the television sensor toward or away from the focal point of the objective lens to produce the sharpest image.

Foot-candle fc   A unit of illuminance on a surface 1 square foot in area on which there is incident light of 1 lumen. The illuminance of a surface placed 1 foot from a light source that has a luminous intensity of 1 candle.

Foot-lambert   A measure of reflected light in a 1 ft. area. A unit of luminance equal to 1 candela per square foot or to the uniform luminance at a perfectly diffusing surface emitting or reflecting light at the rate of 1 lumen per square foot.

FOV Field of view   The width, height, or diameter of a scene to be monitored, determined by the lens focal length, the sensor size, and the lens-to-subject distance. The maximum angle of view that can be seen through a lens or optical assembly. Usually described in degrees, for a horizontal, vertical, or circular dimension.

Frame   A frame is a complete video picture made up of two separate fields of 262.5 lines (NTSC) and 312.5 lines in (CCIR). In the standard U.S. NTSC 525-line system, the frame time is 1/30 second. In the European 625-line system, the frame time is 1/25 second. In a camera with progressive scan, each frame is scanned line-by-line and not interlaced.

Frame frequency   The number of times per second that the frame is scanned. The U.S. NTSC standard is 30 times per second. The European standard is 25 times per second.

Frequency interlace   The method by which color and black-and-white sideband signals are interwoven within the same channel bandwidth.

Frequency response   The range or band of frequencies to which a unit of electronic equipment will offer essentially the same characteristics.

Front (First) surface mirror   An optical mirror with the reflecting surface applied to the front surface of the glass instead of to the back. The first surface mirror is used to avoid ghost images. In the more common rear surface mirror the light has to first pass through the mirror glass, strike the rear surface and then exit back out through the front of the mirror. This causes a secondary image or ghost image. The reflecting material on first surface mirrors is usually aluminum with a silicon monoxide protective overcoat.

Front porch   That portion of a composite picture signal which lies between the leading edge of the horizontal blanking pulse and the leading edge of the corresponding sync pulse.

f-stop   see **f-number**.

FTP File Transfer Protocol   A part of the primary Internet protocol group TCP/IP is the simplest way to transfer files between computers from the Internet servers to the client computer. Like HTTP which transfers displayable web pages and related files, and which transfers e-mail, FTP is an application protocol that uses the Internet's TCP/IP protocols.

Gain   An increase in voltage, current, power, or light usually expressed as a positive number or in decibels.

GaAs Gallium arsenide diode   A light-emitting diode (LED) semiconductor device that emits low-power infrared radiation. Used in television systems for covert area illumination or with fiber optics for signal transmission. The radiation is incoherent and has a typical beam spread of 10 to 50 degrees and radiates at 850 or 960 nanometers in the IR spectrum.

Gallium arsenide laser   A narrow-band, narrow-beam IR radiation device. The radiation is coherent, has a very narrow beam pattern, typically less than 1/2 to 2 degrees, and radiates in the IR spectrum.

Galvanometer   A device that converts an electrical signal into mechanical movement without the complexity of a DC motor. Used to control the iris vanes in the camera lens.

Gamma   A numerical value of the degree of contrast in a television picture that is used to approximate the curve of output magnitude versus input magnitude over the region of interest. Gamma values range from 0.6 to 1.0.

Gamma correction   To provide for a linear transfer characteristic from input to output device by adjusting the gamma.

Genlock   An electronic device used to lock the frequency of an internal sync generator to an external source.

Geometric distortion   Any aberration that causes the reproduced picture to be geometrically dissimilar to the original scene.

Ghost   A spurious image resulting from an echo (electrical) or a second or multiple reflection (optical). A front surface mirror produces no ghost, while a rear surface mirror produces a ghost. Airwave RF and microwave signals reaching a receiver after reflecting from multiple paths produce ghosts.

GIF Graphics Interchange Format   One of the two most common file formats for graphic images on the World Wide Web. The other is JPEG.

Gigabit Ethernet The latest version of Ethernet offering 1000 Mbps or 1 gigabit per second (Gbps) bandwidth that is 100 times faster than the original Ethernet. It is compatible with existing Ethernets since it uses the same CSMA/CD and MAC protocols.

Grayscale Variations in value from white, through shades of gray, to black, on a television screen. The gradations approximate the tonal values of the original image picked up by the TV camera. Most analog video cameras produce at least 10 shades of gray. Digital cameras typically produce 256 levels of gray.

GUI Graphic User Interface A digital user control and processing for the user of that system. The Macintosh or Microsoft Windows operating systems are examples of GUI systems.

H.264 A powerful MPEG compression algorithm standard developed through the combined effort of the ITU and MPEG organizations providing excellent compression efficiency and motion detection attributes. See **MPEG**.

Halo A glow or diffusion that surrounds a bright spot on a television picture tube screen or image intensifier tube.

Hertz (Hz) The frequency of an alternating signal formerly called cycles per second. The U.S. in Japan power-line frequency is 60 Hz. Most European countries use a 50 Hz power line frequency.

High-contrast image A picture in which strong contrast between light and dark areas is visible and where intermediate values, however, may be missing.

High Definition Television HDTV A television standard using digital signals. HDTV signals contain over 720 TV lines of resolution compared with 525 TV line (NTSC) and 625 TV line (PAL) in legacy analog standards. HDTV video formats generally use a 1080i or 720p image format and have a 16:9 aspect ratio.

High-frequency distortion   Distortion effects that occur at high frequency. In television, generally considered as any frequency above 15.75 KHz.

Highlights   The maximum brightness of the TV picture occurring in regions of highest illumination.

Horizontal blanking   Blanking of the picture during the period of horizontal retrace.

Horizontal hum bars   Relatively broad horizontal bars, alternately black and white, that extend over the entire picture. They may be stationary or may move up or down. Sometimes referred to as a "venetian-blind" effect. In 60-Hz systems, hum bars are caused by approximate 60-Hz interfering frequency or one of its harmonic frequencies (such as 120 Hz).

Horizontal retrace   The return of the electron beam from the right to the left side of the raster after scanning one horizontal line.

Horizontal, vertical resolution   see **Resolution**.

HTML Hypertext Markup Language   A simple document formatting language used for preparing documents to be viewed by a tool such as a WWW browser. A set of markup symbols or codes inserted in a file posted on the WWW browser. HTML instructs the web browser how to display web pages and images.

HTTP Hypertext Transfer Protocol   A set of rules for exchanging files that governs transmission of formatted documents (text, graphic images, sound, video, and other files) for viewing over the WWW and Internet. HTTP is an application protocol relative to the TCP/IP suite of protocols, which are the basis for information exchange on the Internet.

Hub   A networking device that enables attached devices to receive data streams that are transmitted over a network and interconnects clients and servers. This device makes it possible for devices to share the network bandwidth available on a network. In data communications a hub is a place of convergence where data arrives from one or more locations and is forwarded to one or more other locations. The hub acts as a wiring concentrator in networks based on *star* topologies, rather than bus topologies in which computers are *daisy-chained* together. A hub usually includes a switch, which is also sometimes considered a hub. With respect to its switching capability, a hub can also be considered a router.

Hue   Corresponds to colors such as red, blue, and so on. Black, gray, and white do not have hue.

Hum   Electrical disturbance at the power supply frequency or harmonics thereof.

Hum modulation   Modulation of a radio frequency, video, or detected signal, by hum.

ICCD Intensified CCD   A charge coupled device sensor camera, fiber optically coupled to an image intensifier. The intensifier is a tube or microchannel plate.

IEC International Telecommunications Union   An international organization that sets standards for the goods and services in electrical and electronic engineering. See **ISO**.

IDE Integrated Drive Electronics   A hard disk drive with built-in electronics necessary for use with a computer. A popular interface to attach hard drives to PCs where the electronics of the controller are integrated with the drive instead of on a separate PC card.

Identification, image   In television, the criterion used to determine whether an object or person can be identified in the scene. It requires approximately 7 TV-line pairs to identify an object or person.

IEEE Institute of Electronic and Electrical Engineers   A technical organization writing standards and publishing technical articles for the electronic and electrical industry.

Illuminance   Luminous flux incident per unit area of a surface; luminous incidence.

Illumination, direct   The lighting produced by visible radiation that travels from the light source to the object without reflection.

Illumination, indirect   The light formed by visible radiation that, in traveling from the light source to the object, undergoes one or more reflections.

Image   A reproduction of an object produced by light rays. An image-forming optical system collects light diverging from an object point and transforms it into a beam that converges toward another point. Transforming all the points produces an image.

Image distance   The axial distance measured from the image to the second principal point of a lens.

Image format   In television, the size of the area of the image at the focal plane of a lens, which is scanned by the video sensor.

Image intensifier   A class of electronic imaging tubes equipped with a light-sensitive photocathode—electron emitter at one end, and a phosphor screen at the other end for visual viewing. An electron tube or microchannel plate (MCP) amplifying (intensifying) mechanism produces an image at its output brighter than the input. The intensifier can be coupled by fiber optics or lenses to a CCD or CMOS sensor. The intensifier can be single stage or multistage, tube or MCP.

Image pickup tube   An electron tube that reproduces an image on its fluorescent screen of an irradiation pattern incident on its input photosensitive surface.

Image plane   The plane at right angles to the optical axis at the image point.

Impedance   The input or output characteristic of an electrical system or component. For maximum power and signal transfer, a cable used to connect two systems or components must have the same characteristic impedance as the system or component. Impedance is expressed in ohms. Video distribution systems have standardized on 75-ohm unbalanced and 124-ohm balanced coaxial cable. UPT uses a 100 ohm impedance. RF and microwave systems use 50-ohm impedance coax.

Incident light   The light that falls directly onto an object.

Infrared radiation   The invisible portion of the electromagnetic spectrum that lies beyond about 750 nanometers (red end of the visible spectrum) and extends out to the microwave spectrum.

Interference   Extraneous energy that interferes and degrades the desired signal.

Interlace, 2 to 1   A scanning format used in video systems in which the two fields comprising the frame are synchronized precisely in a 2 to 1 ratio, and where the time or phase relationship between adjacent lines in successive fields is *fixed*.

Interlace, random   A scanning technique used in some systems in which the two fields making up the frame are not synchronized, and where there is **no** fixed time or phase relationship between adjacent lines in successive fields.

Interlaced scanning   A scanning process used to reduce image flicker and electrical bandwidth. The interlace is 2:1 in the NTSC system.

Internet   A massive global network that interconnects tens of thousands of computers and networks worldwide and is accessible from any computer with a modem or router connection and the appropriate software.

Intranet   A network internal to an organization that takes advantage of some of the same tools popularized on the Internet. These include browsers for viewing material, HTML for preparing company directories or announcements, etc.

IP Address   On the Internet each computer and connected appliance (camera switcher, router, etc.) must have a unique address. This series of numbers functions similarly to a street address, identifying the location of both sender and recipient for information dispatched over the computer network. The IP address has 32 bits in an 8 bit quad format. The four groups in decimal format are separated by a period (.). Two quad groups represent the network and two the machine or host address. An example of an IP address is 124.55.19.64. See **Subnet, Subnet mask**.

IP Internet Protocol   The method by which data is sent from one computer to another over the Internet. Each computer, known as a host on the Internet has one address that uniquely identifies it from all other computers on the Internet. A Web page or an e-mail is sent or received by dividing it into blocks called packets. Each packet contains both the sender's Internet address and the receiver's address. Each of these packets can arrive in an order different from the order from which they were sent. The IP just delivers them and the Transmission Control Protocol TCP puts them in the correct order. The most widely used version of the IP is IP Version 4 (IPv4).

Iris   An adjustable optical-mechanical aperture built into a camera lens to permit control of the amount of light passing through the lens.

Iris diaphragm   A mechanical device within a lens used to control the size of the aperture through which light passes. A device for opening and closing the lens aperture to adjust the f-stop of a lens.

ISC International Security Conference A trade organization to provide a forum for manufactures to display their products. ISC provides accredited security seminars for attendees.

ISDN Integrated Services Digital Network A communication protocol offered by telephone companies that permits high-speed connections between computers and networks in remote locations.

ISIT Intensified silicon intensified target A SIT tube with an additional intensifier, fiber-optically coupled to provide increased sensitivity.

ISO International Organization for Standardization A worldwide federation of national standards bodies from over 130 countries to promote the worldwide standardization of goods and services. ISO's work results in international agreements that are published as international standards. The scope of ISO covers all technical fields except electrical and electronic engineering, which is the responsibility of IEC. Among well-known ISO standards is the ISO 9000 business standard that provides a framework for quality management and quality assurance.

Isolation amplifier An amplifier with input and output circuitry designed to eliminate the effects of changes made by either upon the other.

ISP Internet Service Provider A company or organization that provides Internet access for companies, organizations, or individuals.

ITU International Telecommunications Union An international organization within which governments in the private sector coordinate global telecom networks and services.

Jitter Instability of a signal in either its amplitude, phase, delay, or pulse width due to environmental disturbances or to changes in supply voltage, temperature, component characteristics, etc.

JPEG Joint Photographic Experts Group   A standards group that defined a compression algorithm commonly called JPEG that is used to compress the data in portrait or still video images. The JPEG file format is the ISO standard 10918 that includes 29 distinct coding processes. Not all must be used by the implementer. The JPEG file type used with the GIF format is supported by the WWW protocol, usually with the file suffix "jpg".

Kell factor   The ratio of the vertical resolution to the number of scanning lines. The empirical number that reduces the vertical resolution of television images from the actual number of lines to 0.7 of that number. For the NTSC system the maximum resolution is reduced to approximately 350 lines.

Lag   The persistence of the electrical charge image for two or more frames after excitation is removed and found in an intensifier tube or monitor display.

LAN Local Area Network   A digital network or group of network segments confined to one building or campus. The LAN consists of a series of PCs that have been joined together via cabling so that resources can be shared, including file and print services.

LASER Light Amplification by Stimulated Emission of Radiation   A LASER is an optical cavity, with plane or spherical mirrors at the ends, that is filled with a light-amplifying material, and an electrical or optical means of stimulating (energizing) the material. The light produced by the atoms of the material generates a brilliant beam of light that is emitted through one of the semi-transparent mirrors. The output beam is highly monochromatic (pure color), coherent, and has a narrow beam (small fraction of a degree).

Laser diode   see **Gallium arsenide laser**.

LCD Liquid Crystal Display   A solid-state video display created by sandwiching an electrically reactive substance between two electrodes. LCDs can be darkened or lightened by applying and removing power. Large numbers of LCD pixels group closely together act as pixels in a flat-panel display.

Leading edge   The major portion of the rise of a pulse, waveform taken from the 10 to 90% level of total amplitude.

Lens   A transparent optical component consisting of one or more optical glass elements with surfaces so curved (usually spherical) that they serve to converge or diverge the transmitted rays of an object, thus forming a real or virtual image of that object.

Lens, fresnel   Figuratively a lens that is cut into narrow rings and flattened out. In practice a thin plastic lens that has narrow concentric rings or steps, each acting to focus radiation into an image.

Lens speed   Refers to the ability of a lens to transmit light. Represented as the ratio of the focal length to the diameter of the lens. A fast lens would be rated $f/1.4$. A much slower lens might be designated as $f/8$. The larger the f-number, the slower the lens. See **f-number**.

Lens system   Two or more lenses so arranged as to act in conjunction with one another.

Light   Electromagnetic radiation detectable by the eye, ranging in wavelength from about 400 nm (blue) to 750 nm (red).

Limiting resolution   A measure of resolution usually expressed in terms of the maximum number of TV lines per TV picture height discernible on a test chart.

Line amplifier   An amplifier for audio or video signals that drive a transmission line. An amplifier, generally broadband, installed at an intermediate location in a main cable run, to compensate for signal loss.

Linearity   The state of an output that incrementally changes directly or proportionally as the input changes.

Line pairs   The term used in defining television resolution. One TV line pair constitutes one black line and one white line. The 525 NTSC system has 485 line pairs displayed.

LLL Low light level   Camera and video systems capable of operating below normal visual response. An intensified video camera such as an ICCD capable of operating in extremely poorly lighted areas.

Load   That component which receives the output energy of an electrical device.

Loss   A reduction in signal level or strength, usually expressed in dB. A power dissipation serving no useful purpose.

Lossless Compression   A form of video compression that does not degrade the quality of the image.

Lossy Compression   A form of compression in which image quality is degraded during compression.

Low-frequency distortion   Distortion effects that occur at low frequency. In video, generally considered as any frequency below 15.75 kHz.

Lumen (lm)   The unit of luminous flux, equal to the flux through a unit solid angle (steradian) from a uniform point source of 1 candela or to the flux on a unit surface of which all points are at a unit distance from a uniform point source of 1 candela.

Luminance   A parameter that represents brightness in the video picture. Luminous intensity (photometric brightness) of any surface in a given direction per unit of projected area of the surface as viewed from that direction, measured in foot-lamberts. Abbreviated as Y.

Luminance signal   The part of the NTSC composite color signal that contains the scene brightness or black and white information.

Luminous flux   The time rate of flow of light.

Lux   International system unit of illumination in which the meter is the unit of length. One Lux equals 1 lumen per square meter.

MAC Media Access Control Protocol   The MAC is the physical address for any device used in a network: a computer, router, IP camera, etc. The address consists of two parts and is 6 bytes long. The first 3 bytes identify the company and the last 3 bytes are the device serial number.

Magnetic focusing   A method of focusing an electron beam by the action of a magnetic field.

Magnification   A number expressing the change in object to image size. Usually expressed with a 1-inch focal length lens and a 1-inch format sensor as a reference (magnification = M = 1). A lens with a 2-inch focal length is said to have a magnification of M = 2.

MAN Metropolitan Area Network   A large network usually connected via fiber optics to obtain the Gbit speeds and huge volumes of digital transmission over long distances.

Matching   The obtaining of like electrical impedances to provide a reflection-free transfer of signal.

Matrix switcher   A combination or array of electromechanical or electronic switches that route a number of signal sources to one or more designations. In video, cameras are switched to monitors, recorders and networks.

Maximum aperture   The largest size the iris diaphragm of the lens can be opened resulting in the lowest lens f-number.

Megabits per second Mbps   Defines the speed at which data is traveling and is measured in millions of bits per second. This is a measure of the performance of a device.

Mercury arc lamp   An intense electric arc lamp that generates blue-white light when electric current flows through mercury vapor in the lamp.

Metal arc lamp   An intense arc lamp that generates a white light when an electric current flows through the multimetal vapor in the lamp.

MHz Megahertz   Unit of frequency equal to 1 million Hz.

Microcomputer   A tabletop or portable digital computer composed of a microprocessor, active memory storage, and permanent memory storage (disk) and which computes and controls functions via a software operating system and applications program.

Micron   Unit of length: one millionth of a meter.

Microphonics   Audio-frequency noise caused by the mechanical vibration of elements within a system or component.

Microprocessor   The brain of the microcomputer. A very large scale integrated circuit comprising the computing engine of a microcomputer. The electronic chip (circuit) that does all the calculations and control of data. In larger machines it is called the central processing unit (CPU).

Microwave transmission   In television, a transmission means that converts the camera video signal to a modulated (AM or FM) microwave signal via a transmitter, and a receiver that demodulates the received microwave signal to the baseband CCTV signal for display on a monitor.

Mirror, first or front surface   An optical component on which the reflecting surface is applied to the *front* of the glass instead of the back, the front being the first surface of incidence and reflectance. It produces a single image with no ghost. See **First Surface Mirror**.

Mirror, rear surface   The common mirror in which the reflecting surface is applied to the *rear* of the glass. It produces a secondary or ghost image.

M-JPEG   A digital video compression format developed from JPEG, a compression standard for still images. When JPEG is extended to a sequence of pictures in the video stream it becomes M-JPEG or motion-JPEG.

Modem   Derived from its function: **modulator-demodulator**. A device that enables a computer to connect to other computers and networks using ordinary phone lines. Modems modulate the digital signals of the computer into analog signals for transmission and then demodulate those analog signals back into digital language that the computer on the other end can recognize. See **Codec**.

Modulation   The process, or results of the process, whereby some characteristic of one signal is varied in accordance with another signal. The modulated signal is called the carrier. The carrier may be modulated in several fundamental ways including: by varying the amplitude, called amplitude modulation (AM); by varying the frequency, called frequency modulation (FM); or by varying the phase, called phase modulation (PM).

Moire pattern   The spurious pattern in the reproduced picture resulting from interference beats between two sets of periodic structures in the image. Usually caused by tweed or checkerboard patterns in the scene.

Monitor   A CRT based monochrome or color display for viewing a television picture from a camera output. The monitor does not incorporate a VHF or UHF tuner and channel selector and displays the composite video signal directly from the camera, DVR, VCR, or any special-effects generator. Monitors take the form of a CRT, LCD, plasma, and other.

Monochrome signal   A black and white video signal with all shades of gray. In monochrome television, a signal for controlling the brightness values in the picture. In color television, that part of the signal which has major control of the brightness values of the picture, whether displayed in color or in monochrome. The minimum number of shades of gray for good image rendition is 10.

Monochrome transmission   The transmission of a signal wave that represents the brightness values in the picture, not the color (chrominance) values.

Motion detector   A device used in security systems that reacts to any movement in a CCTV camera image by automatically setting off an alarm and/or indicating the motion on the monitor.

Motorized lens   A camera lens fitted with small electric motors that can focus the lens, open or close the iris diaphragm, or in the case of the zoom lens, change the focal length by remote control.

MPEG-4   A compression standard formulated by the Moving Pictures Experts Group. The MPEG-4 standard for digital video and audio compression is optimized for moving images in which the compression is based on the similarity of successive pictures. MPEG-4 files carry an .mpg suffix. See also **H.264**.

Multicasting   Refers to the propagation from one source to a subset of potential destinations. A technique for simultaneously sending multiple digital video streams on a single channel.

Multiplexer   High speed electronic switch that combines two or more video signals into a single channel to provide full-screen images up to 16 or 32 displayed simultaneously in split image format. Multiplexers can play back everything that happened on any one camera without interference from the other cameras on the system.

NAB National Association of Broadcasters

Nanometer (nm)   Unit of length: one billionth of a meter.

NA Numerical aperture The sine of the half-angle of the widest bundle of rays capable of entering a lens, multiplied by the refractive index of the medium containing that bundle. In air, the refractive index n = 1.

ND Spot Filter A graduated filter at the center of a lens that has minimal effect when the iris is wide open but increases its effect as the iris closes. The filter has a varying density as a function of the distance from the center of the lens with maximum density at the center of the filter.

ND Neutral density filter An optical attenuating device or light filter that reduces the intensity of light without changing the spectral distribution of the light. Can be attached to the lens of the camera to assist in preventing over exposure of an image. See **Density**.

Negative image A picture signal having a polarity that is opposite to normal polarity and that results in a picture in which the roles of white and black areas are reversed.

Network A collection of devices that include computers, printers, and storage devices that are connected together for the purpose of sharing information and resources.

Newvicon A former television pickup tube with a cadmium and zinc telluride target with sensitivity about 20 times that of a vidicon target. It had a spectral response of 470 to 850 nm, good resolution, and was relatively free from burn-in.

NIC Network Interface Card A device that provides for connecting a PC to a network. NIC cards are also called network adapters and provide an essential link between a device and the network.

Noise The word noise originated in audio practice and refers to random spurts of acoustical or electrical energy or interference. In television, it produces a "salt-and-pepper" pattern over the televised picture. Heavy noise is sometimes referred to as "snow."

Non-browning   A term used in connection with lens glass, faceplate glass, fiber optics, and in radiation-tolerant television cameras. Non-browning glass does not discolor (turn brown) when irradiated with atomic particles and waves.

Non-composite video   A video signal containing all information except synchronization pulses.

Notch filter   A special filter designed to reject a very narrow band of electrical frequencies or optical wavelengths.

NTSC National Television Systems Committee   The committee that worked with the FCC in formulating standards for the original monochrome and present-day U.S. color television system. NTSC has 525 horizontal scan lines, 30 frames per second, and a bandwidth of 4.2 MHz. NTSC uses a 3.579545 MHz color sub-carrier. It employs 525 lines per frame, 29.97 frames/sec and 59.94 fields/sec. The NTSC standard is used in the United States and Japan.

NVR Network Video Recorder   A software or computer that records video on a hard disk. Like a DVR, it records digitally so the user can instantly search by time, date, and camera. It collects video from network cameras, network video servers, or a DVR over the network.

Object distance   The distance between the object and the cornea of the eye, or the first principal point of the objective in an optical device.

Objective lens   The optical element that receives light from an object scene and forms the first or primary image. In cameras, the image produced by the objective is the final image. In telescopes and microscopes, when used visually, the image formed by the objective is magnified by an eyepiece.

Optical axis   The line passing through the centers of curvatures of the optical surfaces of a lens or the geometric center of a mirror or window; the optical centerline.

Optical splitter   An optical lens-prism and/or mirror system that combines two or more scenes and images them onto one video camera. Only optical components are used to combine the scene.

Optical zoom   Optical zoom is produced by the lens itself by moving sets of lenses in the zoom lens to provide a continuous, smooth change in focal length from wide-angle to narrow-angle (telephoto).

Orientation, image   In television, the criterion used to determine the angular orientation of a target (object, person) in an image. At least 2 TV-line pairs are required.

Overshoot   The initial transient response to a unidirectional change in input, which exceeds the steady-state response.

Overt surveillance   In television, the use of any openly displayed television lenses or cameras to view a scene.

Packet   A block of data with a "header" attached that can indicate what the packet contains and where it is headed. A packet is a "data envelope" with the header acting as an address.

PAL Phase Alternating Line   A European color television system using a 625 lines per frame 25 frames/second composite analog color video system at 5.5 MHz bandwidth. In this system the sub-carrier derived from the color burst is inverted in phase from one line to the next in order to minimize errors in hue that may occur in color transmission. The PAL format is used in Western Europe, Australia, parts of Africa, and the Middle East.

Pan and tilt   Camera-mounting platform that allows movement in both the azimuth (pan) and the elevation (tilt) planes.

Pan, panning   Rotating or scanning a camera around a vertical axis to view an area in a horizontal direction.

Pan/tilt/zoom   Three terms associated with television cameras, lenses, and mounting platforms to indicate the horizontal (pan), vertical (tilt), and magnification (zoom) they are capable of producing.

Passive   In video, cameras using ambient visible or IR, contrasted to using an active IR illuminator. In electronics, a non-powered device that generally presents some loss to a system and is incapable of generating power or amplification.

Peak-to-peak   The amplitude (voltage) difference between the most positive and the most negative excursions (peaks) of an electrical signal.

Pedestal level   see **Blanking level**.

Persistence   In a cathode ray tube, the period of time a phosphor continues to glow after excitation is removed.

Phased array antenna   A transmit or receive antenna comprised of multiple identical radiating elements in a regular arrangement and fed or connected to obtain a prescribed radiation pattern.

Phosphor   A substance capable of luminescence used in fluorescent lamps, television monitors, viewfinders, and image intensifier screens.

Phosphor-dot faceplate   A glass plate in a tri-color picture tube. May be the front face of the tube or a separate internal plate. Its rear surface is covered with an orderly array of tri-color lines or phosphor dots. When excited by electron beams in proper sequence, the phosphors glow in red, green, and blue to produce a full-color picture.

Photocathode   An electrode used for obtaining photoelectric emission.

Photoconductivity   The changes in the electrical conductivity (reciprocal of resistance) of a material as a result of absorption of photons (light).

Photoconductor   A material whose electrical resistance varies in relationship with exposure to light.

Photoelectric emission   The phenomenon of emission of electrons by certain materials upon exposure to radiation in and near the visible region of the spectrum.

Photon-limited sensitivity   When the quantity of available light is the limiting factor in the sensitivity of a device.

Photopic vision   Vision that occurs at moderate and high levels of luminance and permits distinction of colors. This light-adapted vision is attributed to the retinal cones in the eye. In contrast, twilight or scotopic vision uses primarily the rods responding to overall light level.

Pickup tube   A television camera image pickup tube. See **Image pickup tube**.

Picture size   The useful area of an image sensor or display. In the standard NTSC format, the horizontal to vertical ratio is 4:3. The diagonal is units. The HDTV ratio is 16:9.

Picture tube   see **Cathode ray tube**.

Pin-cushion distortion   Distortion in a television picture that makes all sides appear to bulge inward.

Pinhole lens   A lens designed to have a relatively small (0.06 inch to 0.375 inch) front lens diameter to permit its use in covert (hidden) camera applications.

PIP Picture in a picture   A video display mode which puts several complete video images on the screen at the same time. Most common is one small image into a large image.

Pixel   Short for Picture element. Any segment of a scanning line, the dimension of which along the line is exactly equal to the nominal line width. A single imaging unit that can be identified by a computer.

Pixelization   An effect seen when an image is enlarged (electronically zoomed) too much and the pixels become visible to the eye.

POTS Plain Old Telephone Service   The original and slowest telephone system.

Preamplifier   An amplifier used to increase the output of a low-level source so that the signal can be further processed without additional deterioration of the signal-to-noise ratio.

Preset   A term used in television pointing systems (pan/tilt/zoom). A computer stores pre-entered azimuth, elevation, zoom (magnification), focus, and iris combinations, which are later accessed when commanded by an operator or automatically on alarm.

Primary colors   Three colors wherein no mixture of any two can produce the third. In color television these are the additive primary colors red, green, and blue (R,G,B).

Progressive or sequential scan   A method of image scanning that processes image data one line of pixels at a time. Each frame is composed of a single field. Contrasted to interlace scanning having two fields per frame.

PSTN Public Switched Telephone Network   The traditional, wired telephone network.

Pulse   A variation of a quantity whose value is normally constant. This variation is characterized by a rise and decay and has finite amplitude and duration.

Pulse rise-time   Time interval between upper and lower limits of instantaneous signal amplitude; specifically, 10 and 90% of the peak-pulse amplitude, unless otherwise stated.

Quad An electronic device having four camera inputs that can display the four cameras simultaneously in a quad format, singly full screen, or full screen sequentially. Alarm input contacts are provided so that the unit switches from a quad display to a full screen image of the alarmed camera.

Radiation Pattern A graphical representation in either polar or rectangular coordinates of the spatial energy distributions of an antenna.

RAID Redundant Array of Independent Disks A system in which a number of hard drives are connected into one large segmented mass storage device. There are RAID-0 to RAID 6 systems.

RAM Random Access Memory The location in the computer where the operating system, application programs, and data in current use are temporarily kept so that they can be quickly reached by the computers processor. RAM is volatile memory, meaning that when the computer is turned off, crashes, or loses power, the contents of the memory are lost.

Random interlace see **Interlace, random**.

Raster The blank white screen that results from the scanning action of the electron beam in a CRT with no video picture information applied. A predetermined pattern of scanning lines that provides substantially uniform coverage of an area. The area of a camera or CRT tube scanned by the electron beam.

Raster burn see **Burn-in**.

Recognition, image In television, the criterion used to determine whether an object or person can be recognized in a television scene. A minimum of 5 TV-line pairs are required to recognize a person or object.

Reference black level The picture signal level corresponding to a specified maximum limit for black peaks.

Reference white level   The picture signal level corresponding to a specified maximum limit for white peaks.

Resolution   A measure of how clear and sharp a video image is displayed on a monitor. The more pixels, the higher the resolution. It measures picture details that can be distinguished on the television screen. Vertical resolution refers to the number of *horizontal* black-and-white or color lines that can be resolved in the picture height. Horizontal resolution refers to the number of *vertical* black-and-white or color lines that can be resolved in a picture width equal to the picture height.

Resolution, horizontal   The amount of resolvable detail in the horizontal direction in a picture; the maximum number of individual picture elements that can be distinguished. It is usually expressed as a number of distinct vertical lines, alternately black and white, that can be seen in a distance equal to picture height. 500 to 600 TV lines are typical with the standard 4.2-MHz CCTV bandwidth. In analog video systems the horizontal resolution is dependent on the system bandwidth.

Resolution, vertical   The amount of resolvable detail in the vertical direction in a picture. It is usually expressed as the number of distinct horizontal lines, alternately black and white, which can be seen in a picture. 350 TV lines are typical in the 525 NTSC system.

Retained image   Also called image burn. A change produced in or on the sensor target that remains on the output device (such as a CRT) for a large number of frames after the removal of a previously stationary light image.

RF Radio frequency   A frequency at which coherent electromagnetic radiation of energy is useful for communication purposes. The entire range of such frequencies, including the AM and FM radio spectrum and the VHF and UHF television spectrum.

RGB Red, Green, and Blue   Abbreviations for the three primary colors captured by a CCD or CMOS imager and displayed in analog and digital video systems. Specifically, the CCD or CMOS camera sensors and CRT, LCD, and plasma displays use RGB resolution elements or pixels. The color signals are mixed electronically to create all the other colors in the spectrum.

Right-angle lens   A multi-element optical component that causes the optical axis of the incoming radiation (from a scene or image focal plane) to be redirected by 90 degrees. It is used when a wide-angle lens is necessary to view a scene at right angles to the camera.

Ringing   In electronic circuits, an oscillatory transient occurring in the output of a system as a result of a sudden change in input.

Ripple   Amplitude variations in the output voltage of a power supply caused by insufficient filtering.

Roll   A loss of vertical synchronization that causes the displayed video image to move (roll) up or down on a television receiver or video monitor.

Roll off   A gradual decrease or attenuation of a signal voltage as a function of frequency.

Router   A device that moves data between different network segments and can look into a packet header to determine the best path for the packet to travel. On the Internet, a device or in some cases software in a computer that determines the next network point to which a data packet should be forwarded towards its final destination. The router analyzes the network status and chooses the optimum path to send each information packet. The router can be located at any juncture of a network or gateway including any Internet access point. The router creates and maintains a table of available network routes and their status. Using distance and cost algorithms it determines the best route for any given packet. Routers allow all users in a network to share a single connection to the Internet or a WAN.

RS170 The original EIA broadcast studio standard issue in November, 1953 for the NTSC black-and-white video format. It described a 2:1 interlaced, 525 line TV standard with the total number of lines occurring in 1/30 second. The vertical frequency was 60 hertz, signal amplitude 1.4 volts peak to peak including sync signal, and bandwidth from 30 Hz to 4.2 MHz.

RS170A The proposed standard for the NTSC composite color video system. Its contents were used in the television industry as a reference, but the document was never adopted. The current color standard is SMPTE 170-1999.

RS232, RS232C A low speed protocol established by the EIA. The standard describes the physical interface and protocol between computers and related devices (printers, modems, etc.). The PC contains a universal asynchronous receiver-transmitter (UART) chip that converts the parallel computer data into serial data for RS232 transmission. The standard recommends a maximum range of 50 feet (15.2 meters) and maximum baud rate of 20 Kbps. There is a standard pinout and the connectors used are D-9 and D-25.

RS422 A protocol established by the EIA consisting of a differential pair of conductors and specified pinout or connectors. The differential pair is one signal transmitted across two separate wires in opposite states: one is inverted and the other is non-inverted. In this differential signal transmission when both lines are exposed to external noise, both lines are affected equally and cancel out. RS422 is usually used in full duplex, four wire mode for point-to-point communication but one transmitter can drive up to 10 receivers. Maximum recommended range and baud rate are 4000 feet and 10Mbps, respectively.

RS485 This protocol is an upgraded version of the RS422 and can handle up to 32 transmitters and receivers by using tri-state drivers. Maximum recommended range and baud rate are 4000 feet and 10 Mbps, respectively.

Saturation   In color, the degree to which a color is undiluted with white light or is pure. The vividness of a color, described by such terms as bright, deep, pastel, pale, and so on. Saturation is directly related to the amplitude of the television chrominance signal.

SCADA Supervisory Control and Data Acquisition   An industrial measurement and control system consisting of a central host or master terminal unit (MTU), one or more field data gathering and control units or remotes, and a collection of standard and/or custom software used to monitor and control remotely located field data elements.

Scanning   Moving the electron beam of an image pickup or a CRT picture tube horizontally across and slowly down the target or screen area, respectively. Moving the charge packets out of a CCD, CMOS, or IR sensor.

Scotopic vision   Vision that occurs in faint light or in dark adaptation. It is attributed to the operation of the retinal rods in the eye and contrasted with daylight or photopic vision, using primarily the cones.

SCSI Small Computer System Interface   A high speed input/output bus that is faster than serial and parallel ports, but slower and harder to configure than USB and FireWire ports. SCSI enables a computer to interact with external peripheral hardware, such as CD-ROM drives, printers, and scanners. SCSI is being supplanted by the newer USB standard.

SDTV Standard Definition Television   Used to describe our 525 line and 625 line interlaced television systems as they are used in the context of DTV.

SECAM Séquential Couleur À Mémorie   A color television system developed in France and used in some countries that do not use either the NTSC or PAL systems. Like PAL, SECAM has

625 horizontal scan lines and 25 frames per second but differs significantly in the method of producing color signals.

Sensitivity   In television, a factor expressing the incident illumination upon a specified scene to produce a specified picture signal at the video camera output.

Server   A computer or software program that provides services to other computer programs in the same computer or other computers. When a computer runs a server program, the computer is called a server. When the server is connected to the Web the server program serves the requested HTML pages or files to the client. Web browsers are clients that request HTML files from Web servers. A server provides services to clients such as: files storage (file server), programs (application server), printer sharing (printer's server), fax or modem sharing.

Set Top Box   A unit similar to a cable box capable of receiving and decoding DTV broadcasts.

Sharpness   Refers to the ability to see the greatest detail in video monitor picture. Color sets marketed before the mid 1980s had sharpness controls to optimize the fine detail in the picture. The use of comb filters in present monitors eliminate the need for the sharpness control. See **Resolution.**

Shutter   In an optical system, an opaque material placed in front of a lens, optical system, or sensor for the purpose of protecting the sensor from bright light sources, or for timing the length of time the light source reaches the sensor or film.

Shuttering   An electronic technique used in solid-state cameras to reduce the charge accumulation from scene illumination for the purpose of increasing the dynamic range of the camera sensor. Analogous to the electronic shutter in a film camera.

Signal strength   The intensity of the video signal measured in volts, millivolts, microvolts, or decibels. Using 0 dB as the standard reference is equal to 1000 microvolts (1 millivolt) in RF systems, and 1 volt in video systems.

Signal-to-noise ratio S:R, S/N   The ratio of the peak value of the video signal to the value of the noise. Usually expressed in decibels. The ratio between a useful television signal and disturbing, unwanted image noise or "snow".

Silicon monoxide   A thin-film dielectric (insulator) used as a protective layer on aluminized mirrors. It is evaporated onto the mirror as a thin layer, and after exposure to the air the monoxide tends to become silicon dioxide or quartz, which is very hard and completely transparent.

Silicon target tube   The successor to the vidicon. A high-sensitivity television image pickup tube of the direct-readout type using a silicon diode target made up of a mosaic of light-sensitive silicon material. It has a sensitivity between 10 and 100 times more than a sulfide vidicon and has high resistance to image burn-in.

SIT Silicon intensified target   A predecessor to the ICCD. An image intensifier fiber-optically coupled to a silicon faceplate resulting in a sensitivity 500 times that of a standard vidicon.

Slow-scan   A first generation video transmission system consisting of a transmitter and receiver to transmit single frame video images at rates slower than the normal NTSC frame rate of 30 per second. The CCTV frames were modulated and transmitted over the phone lines to a distant receiver-demodulator and displayed on a CCTV monitor. The slow-scan process periodically sends "snapshots" of the scene with typical sending rates of 1 to 5 frames per second.

Sodium lamp   A low or high-pressure discharge metal vapor arc lamp using sodium as the luminous radiation source. The lamp produces a yellow light and has the highest electrical-to-light output efficiency (efficacy) of any lamp. Because of their poor color balance neither is recommended for color CCTV systems, but can be used for monochrome systems.

SMPTE Society of Motion Picture and Television Engineers A global organization based in the U.S. that sets standards for baseband visual communications.

SMTP Simple Mail Transfer Protocol A TCP/IP that is used for sending and receiving e-mail. To improve its usefulness it is used with the POP3 or IMAP protocols, allowing the user to save messages in a server mailbox and download them periodically from the server. Users use the SMTP to send messages and POP3 or IMAP to receive messages from the local server.

Snow Heavy random noise manifest on a phosphor screen as a changing black and white or colored "peppered" random noise. See **Noise**.

Speckle Noise manifest in image intensifiers in the form of small, localized bright light spots or flashes seen in the device monitor.

Spike A transient of short duration, comprising part of a pulse, during which the amplitude considerably exceeds the average amplitude of the pulse.

Spread Spectrum Modulation SSM A communication technique that spreads a signal bandwidth over a wide range of frequencies for transmission and then de-spreads it to the original data bandwidth at the receiver. See **Frequency Hopping**.

Subnet A uniquely identifiable part of a network. The subnet may represent a particular department at a location or a particular geographical location of the subnet in a building on the local area network (LAN). Dividing the network into sub-networks allows it to be connected to the Internet with a single shared network address. See **IP address**.

Subnet mask This set of numbers tells a signal router which numbers are relevant under the IP address mask. In the binary mask system a "1" over a number indicates that the number under the 1 is relevant, and a "0" over a number says ignore the number under it. Using the mask means the router does not have to look at all 32-bits in the IP address. See **IP address**.

Super VGA   A video format providing high-quality analog video by separating the video signal into three color signals, R, G, and B, allowing for exceptionally clear and bright images.

S-VHS Super VHS   A video tape format in which the chrominance and luminance signals are recorded and played back separately providing for better picture quality.

S-Video   An encoded video signal that separates the luminance (brightness) part of the signal from the chrominance (color) to provide better picture quality.

Switch   A network device that improves network performance by segmenting the network and reducing competition for bandwidth. The switch selects a path or circuit for sending a packet of data to its next destination. When a switch port receives data packets, it forwards those packets only to the appropriate port for the intended recipient. A switch can also provide the function of routing the data packet to the next point in the network. The switcher is faster than the router and can more effectively determine the route the data takes.

Switcher   A video electronic device that connects one of many input cameras to one or several output monitors, recorders, etc., by means of a panel switch or electronic input signal.

Switcher, alarming   An automatic switcher that is activated by a variety of sensing devices. Once activated, the switcher connects the camera to the output device (monitor, recorder, etc.).

Switcher, bridging   A sequential switcher with separate outputs for two monitors, one for a programmed sequence and the second for extended display of a single camera scene.

Switcher, homing   A switcher in which: 1) the outputs of multiple cameras can be switched sequentially to a monitor, 2) one or more cameras can be bypassed (not displayed), or 3) any one of the cameras can be selected for continuous display on the monitor (homing). The switcher has three front-panel controllable modes:

1) Skip, 2) Automatic (sequential) and 3) Select (display one camera continuously). The lengths of time each camera output is displayed are independently selectable by the operator.

Switcher, manual   A switcher in which the individual cameras are chosen by the operator manually by pushing the switch for the camera output signal chosen to be displayed, recorded, or printed.

Switcher, sequential   A generic switcher type that allows the video signals from multiple cameras to be displayed, recorded, or printed one at a time in sequence.

Sync   A contraction of synchronous or synchronization.

Sync generator   A device for generating a synchronizing signal.

Synchronizing   Maintaining two or more scanning processes or signals in phase.

Sync level   The level of the peaks of the synchronizing signal.

Sync signal   The signal employed for the synchronizing of scanning.

Talk-back   A voice inter-communicator; an intercom.

Target   In solid-state sensors, a semiconductor structure using picture elements to accumulate the picture charge, and a scanning readout mechanism to generate the video signal. In surveillance, an object (person, vehicle, etc.) or activity of interest present in an image of the scene under observation. In image pickup tubes, a structure using a storage surface that is scanned by an electron beam to generate a signal output current corresponding to a charge-density pattern stored on it.

TCP Transmission control protocol   A protocol used along with the IP to send data in the form of message units between computers, over the Internet. While IP takes care of handling the actual delivery of the data, TCP takes care of keeping track of the individual units of data (called packets) that a message is divided into, for efficient routing over the Internet.

TCP/IP Transmission control protocol/Internet protocol   The basic communication language (or protocol) of the Internet. It is also used as a communications protocol in private networks called intranets, and extranets. TCP/IP communication is primarily point-to-point in which each communication is from one point (or host computer) to another. The TCP of TCP/IP handles the tracking of the data packets.

TDMA Time Division Multiple Access   A digital multiplexing (channel sharing) technique whereby each signal is sent at a repeating time slot in a frequency channel. Because the data from each user always appears in the same time slot, the receiver can separate the signals.

Telephoto lens   A long focal length lens, producing a narrow field of view. Telephoto lenses are used to magnifier objects within their field of view.

Test pattern   A chart especially prepared for checking overall performance of a television system. It contains combinations of lines and geometric shapes of specific sizes and spacings. In use, the camera is focused on the chart, and the pattern is viewed at the monitor for image fidelity (resolution). The chart most commonly used is the EIA resolution chart.

TFT LCD Thin film transistor LCD   A type of LCD flat-panel display screen. The TFT technology provides the best resolution of all the flat-panel techniques and is also the most expensive. TFT screens are sometimes called active-matrix LCDs.

Tilt   A low frequency signal distortion. A deviation from the ideal low-frequency response. Example: Instead of a square wave having a constant amplitude, it has a tilt.

Time lapse   Capturing a series of images at preset intervals.

Time lapse recorder   The video cassette recorder extends the elapsed time over which it records by recording user selected samples of the video fields or frames instead of recording in real-time. For example, recording every other field produces a 15 field/sec recording and doubles the elapsed time recorded on the tape. Recording every 30th field produces a 1 field/sec recording and provides 30 times the elapsed recording time.

Token ring   LAN technology in which packets are conveyed between network end stations by a "token" moving continuously around a closed ring between all the stations. Operates at 4 or 16 Mbps.

Transient   An unwanted signal existing for a brief period of time that is superimposed on a signal or power line voltage.

Triaxial cable   A double shielded cable construction having a conductor and two isolated braid shields both insulated from each other. The second braid is applied over an inner jacket and an outer jacket applied over the outer braid.

Tri-split lens   A multi-element optical assembly that combines one-third of each of three scenes and brings them to focus (adjacent to one another) at the focal plane of a video camera sensor. Three separate objective lenses are used to focus the scenes onto the splitter assembly.

T-stop   A measurement system used primarily for rating the light throughput of a catadioptric lens having a central obscuration. It provides an equivalent aperture of a lens having 100% transmission efficiency. This system is based on actual light transmission and is considered a more realistic test than the f-stop system.

Tungsten-halogen lamp   An improved tungsten lamp once called quartz-iodine having a tungsten filament and halogen gas in a fused quartz enclosure. The iodine or bromine added to the fill produces a tungsten-halogen cycle that provides self-cleaning and an extended lifetime. The higher filament temperature produces more light and a "whiter" light (higher color temperature).

Tungsten lamp A light source using a tungsten filament surrounded by an inert gas (nitrogen, xenon) enclosed in a glass or quartz envelope. An AC or DC electric current passing though the filament causes it to heat to incandescence, producing visible and infrared radiation.

TV lines A convention used in the video industry to specify the resolution of a video image. Horizontal resolution is measured in TV lines across a width equal to the height of the display. Typical horizontal resolutions in the security industry are 480 TV lines for color and 570 TV lines for monochrome. Vertical resolution is the number of horizontal lines multiplied by the Kell factor.

Twin-lead A transmission line having two parallel conductors separated by insulating material. Line impedance is determined by the diameter and spacing of the conductors and the insulating material and is usually 300 ohms for television receiving antennas.

Twin-split lens A multi-element optical assembly that combines one half of each of two scenes and brings them to focus (adjacent to one another) at the focal plane of a video camera sensor. Two separate objective lenses are used to focus the scenes onto the splitter assembly.

UHF Ultra-high frequency In television, a term used to designate the part of the RF spectrum in which channels 14 through 83 are transmitted. The UHF signals are in the 300 to 3000 MHz frequency range.

UL certified A certification given by Underwriters Laboratory to certain items that are impractical to UL list and which the manufacturer can use to identify the item.

UL listed   A label that signifies that a product meets the safety requirements as set forth by UL safety testing standards.

UL Underwriters Laboratory   A testing laboratory that writes safety standards used by manufacturers when designing products. UL tests and approves manufactured items for certification or listing providing they meet required safety standards.

UNIX   A computer operating system like DOS or MacOS. UNIX is designed to be used by many people at the same time and has TCP/IP built-in. The UNIX operating system was developed by AT& T Bell Labs and was used to develop the Internet.

URL Uniform resource locator   The address of a web site.

USB Universal serial bus   A high-speed port found on most computers that allows a much faster transfer speed than a serial or parallel-port.

UTP Unshielded twisted pair   Two insulated conductors in an insulating jacket in which the two conductors are twisted along the length of the cable. When provided with appropriate transmitter and receiver, UTP provides an alternative to the RG59 coaxial cable.

UV Ultraviolet   An invisible region of the optical spectrum located immediately beyond the violet end of the visible spectrum, and between the wavelengths of approximately 100 and 380 nanometers. Radiation just beyond the visible spectrum (at the blue end of the visible spectrum) ordinarily filtered or blocked to prevent eye damage.

Varifocal Lens   A lens having a manually adjustable focal length providing a range of field of view.

VCR Video cassette recorder   A device that accepts signals from a video camera and a microphone and records images and sound on 1/2" or 1/4" magnetic tape in a cassette. The VCR can play back the recorded images and sound for viewing on a television receiver or CCTV monitor or printing out single frames on a video printer.

Vectorscope   A special oscilloscope used for color camera and color video system calibration. The vectorscope decodes the color information into R-Y and B-Y signals which are used to drive the x and y axis of the scope. The absence of color in the video signal is displayed as a dot at the center of the display. The angle-distance around the circle, and magnitude-distance away from the sensor, indicate the phase and amplitude of the color signal. The vectorscope graphically indicates on a CRT the absolute phase angle between the different color signals with respect to a reference signal, and to each other. These angles represent the phase differences between the signals.

Vertical blanking, Retrace   The process of bringing the scanning electron beam in a CRT from the bottom of the picture back to the top. Vertical retrace occurs between writing each field of a picture. The beam is shut off and blanked during the retrace.

Vertical resolution   The number of horizontal lines that can be seen in the reproduced image of a television pattern. The 525 TV line NTSC system and Kell effect limits the vertical resolution to appproximately 350 TV lines maximum.

Vertical retrace   The return of the electron beam to the top of the picture tube screen or the pickup tube target, at the completion of the field scan. The retrace is not displayed on the monitor.

VGA   A standard display format having an image resolution of 640 × 480 pixels.

VHF Very High Frequency   In television, a term used to designate the part of the RF spectrum in which channels 2 through 13 are transmitted. A signal in the frequency range of from 30 to 300 MHz.

VHS Victor home system   The 1/2″ video tape cassette recording format in most widespread use.

Video   A term pertaining to the bandwidth and spectrum position of the signal resulting from television scanning. In current CCTV usage video means a bandwidth between 30 Hz and 5–6 MHz.

Video amplifier   A wideband amplifier used for amplifying video picture signals.

Video band   The frequency band used to transmit a composite video signal.

Video signal, non-composite   The picture signal. A signal containing visual information without the horizontal and vertical synchronization and blanking pulses. See **Composite video**.

Vidicon tube   An early imaging tube used to convert a visible image into an electrical signal. The spectral response covers most of the visible spectrum and most closely approximates the human eye response (400–700 nm).

Viewfinder   A small electronic or optical viewing device attached to a video camera so that the operator at the camera location can view the scene that the camera sees.

Vignetting   The loss of light through a lens or optical system at the edges of the field due to using an undersized lens or inadequate lens design. Most well-designed lenses minimize vignetting.

Visible spectrum   That portion of the electromagnetic spectrum to which the human eye is sensitive. The range covers from 400 to 700 nanometers.

VPN Virtual Private Network   A private data network that makes use of the public telecommunications infrastructure. The

VPN enables IP traffic to travel securely over a public TCP/IP network by encrypting all traffic from one network to another. The VPN maintains privacy through the use of a "tunneling" protocol and security procedures. It does this at a much lower cost than privately owned or leased lines. Many companies use a VPN for both Extranets and wide-area networks (WAN).

WAN Wide Area Network   A public or private network that provides coverage over a broad geographic area. WANs are typically used for connecting several metropolitan areas as part of a larger network. Universities and large corporations use WANs to connect geographically remote locations.

Waveform monitor   A specialized oscilloscope with controls that allow the display and analysis of analog and digital video waveforms. Parameters analyzed include frequency, waveform shape, presence or absence of synchronizing pulses, etc.

Wavelength   The length of an electromagnetic energy wave measured from any point on one wave to the corresponding point on the next wave, usually measured from crest to crest. Wavelength defines the nature of the various forms of radiant energy that compose the electromagnetic spectrum and determines the color of light. Common units for measurement are the nanometer (1/10,000 micron), micron, millimicron, and the Angstrom.

Wavelet   A unique mathematical function used in signal processing and video image compression. The process is similar to Fourier analysis.

Web browser   A web program that allows Web browsers to retrieve files from computers connected to the Internet. Its main function is to serve pages to other remote computers.

Web server   A program that allows web browsers to retrieve files from computers connected to the Internet. The web server listens for requests from web browsers and upon receiving a request for a file sends it back to the browser.

White clipper   A nonlinear electronic circuit providing linear amplification up to a predetermined voltage and then unity amplification for signals above the predetermined voltage.

White compression   Amplitude compression of the signals corresponding to the white regions of the picture.

White level   The top end of the gray scale. The picture signal level corresponding to a specified maximum limit for white peaks.

White peak   The maximum excursion of the picture signal in the white direction.

White peak clipping   Limiting the amplitude of the picture signal to a pre-selected maximum white level.

Wi-Fi Wireless Fidelity   The Institute of Electrical and Electronic Engineers (IEEE) 802.11 wireless standard for transmitting video images and other data over the airwaves between computers, access points, routers, or other digital video devices.

WLAN Wireless Local Area Network or Wireless LAN   A wireless computer-to-computer data communications network having a nominal range of 1000 ft.

Working distance   The distance between the front surface of an objective lens and the object being viewed.

WWW World Wide Web   The name of the total space for highly graphical and multimedia applications on the Internet.

Xenon arc lamp   An arc lamp containing the rare gas xenon that is excited electrically to emit a brilliant white light. The lamps are available in short-arc (high-pressure) and long-arc (low-pressure).

Yagi Antenna   A multiple element parasitic antenna originated by Yagi-Uda in Japan. A common means of achieving high antenna gain in a compact physical size in the VHF and UHF frequency range. The Yagi antenna consists of a driven element, a reflector element, and one or more director elements.

Y/C   The term used to describe the separate luminance (Y) and chrominance (C) signals. Separating the signals improves the final video image.

Zoom   Optical zooming using a lens to enlarge or reduce the size of the scene image on the video sensor on a continuously variable basis. The wide-angle setting provides low magnification. Narrow-angle (telephoto) setting provides high magnification. Electronic zooming magnifies or de-magnifies the image size of a video scene electronically. All magnification is referenced to the human eye with a magnification of 1.

Zoom lens   A lens capable of providing variable focal lengths. An optical lens system of continuously variable focal length with the focal plane remaining in a fixed position at the camera sensor. Groups of lens components are moved to change their relative physical positions, thereby varying the focal length and angle of view through a specified range of magnifications.

Zoom range   The degree to which the focal length of a camera lens can be adjusted from wide-angle to telephoto. Usually defined with a numerical ratio like 10:1 (telephoto: wide-angle).

Printed and bound by CPI Group (UK) Ltd, Croydon, CR0 4YY

03/10/2024

01040434-0005